PHYSICS RESEARCH AND TECHNOLOGY

PLASMONS

STRUCTURE, PROPERTIES AND APPLICATIONS

PHYSICS RESEARCH AND TECHNOLOGY

Additional books in this series can be found on Nova's website
under the Series tab.

Additional e-books in this series can be found on Nova's website
under the e-book tab.

PHYSICS RESEARCH AND TECHNOLOGY

PLASMONS

STRUCTURE, PROPERTIES AND APPLICATIONS

ANTON E. TURUNEN
AND
JOHANNES O. NIEMI
EDITORS

Nova Science Publishers, Inc.
New York

For permission to use material from this book please contact us:
Telephone 631-231-7269; Fax 631-231-8175
Web Site: http://www.novapublishers.com

LIBRARY OF CONGRESS CATALOGING-IN-PUBLICATION DATA

Plasmons : structure, properties, and applications / editors, Anton E. Turunen and Johannes O. Niemi.
p. cm.
Includes bibliographical references and index.
ISBN 978-1-62100-060-0 (hardcover)
1. Plasmons (Physics) I. Turunen, Anton E. II. Niemi, Johannes O.
QC176.8.P55P5775 2011 530.4'16--dc23 2011030090

Published by Nova Science Publishers, Inc. † New York

CONTENTS

PREFACE

In physics, a plasmon is a quantum of plasma oscillation. The plasmon is a quasiparticle resulting from the quantization of plasma oscillations just as photons and phonons are quantizations of light and mechanical vibrations, respectively. In this book, the authors present current research in the study of the structure, properties and applications of plasmons. Topics discussed include plasmonic biosensors for sensitive measurements in complex biological matrices; generation of surface plasmon resonance for sub-wavelength nano-photolithography; applications of gold nanorods based on localized surface plasmon resonance; the study of living cells by infrared surface plasmon spectroscopy and interface plasmon polaritons in uniaxial bianisotropic structures.

Chapter 1 - Biomolecules of medical or physiological significance are often present only in very low concentrations in biological fluids such as blood, urine or saliva, making them difficult to measure. Such complex matrices pose particular challenges, such as fouling of sensor surfaces, obstruction of specific binding interactions, such as that between an antibody and its corresponding antigen, and production of non-specific binding responses in optical biosensors. Overcoming such challenges is central to the future usefulness of optical biosensor technology in fields such as medical diagnosis, physiological monitoring and the assessment of human health and well-being, particularly in non-invasive techniques where analyte concentrations are typically very low. Plasmons have been very widely utilized in transduction of biochemical binding events on sensing surfaces in the form of surface plasmon resonance (SPR) biosensors. Very sensitive assays can result when such a transduction is coupled with immunoassays, immunomimetic systems such as molecularly imprinted polymers, or complementary DNA binding interactions. These assays are capable of measurement at physiologically relevant concentrations in the blood and sometimes even in non-invasively obtained fluids such as saliva. This chapter will examine the cutting-edge approaches that have been taken in adapting plasmonic biosensors to function in the presence of complex biological matrices. Most of the work in this area has focused on measurement in human blood but there will also be a discussion of emerging techniques for non-invasive measurement using fluids such as saliva and urine. The physicochemical challenges posed by these matrices will be outlined and related to the sophisticated chemical architectures that have been developed to minimize non-specific binding and ideally to help to ensure optimal sensor signal sensitivity. The functioning of the strategies applied will be related to surface chemistry and plasmonic theory. Included in this analysis are the varied physical separation and purification technologies used, where these are integrated with the sensor in an on-line approach. The analytes examined will include proteins, DNA and small molecules. Whilst the predominant emphasis will be on immunoassays, alternative binding interactions such as DNA binding and immunomimetic approaches will also be examined.

Chapter 2 - This chapter discusses the design and construction of surface-enhanced Raman scattering (SERS) active nanostructures with tunable size-/shape- dependent physical and chemical properties, and provides examples of exploit their wide applications in many scientific and technological fields. Coupled with biomolecules, different hybrid devices were developed, which offered a powerful platform to study the conformation change, dynamic behavior and biocatalysis of the target redox proteins like cytochrome c (Cyt-c, usually as a model protein). It would be benefit to deeply understand the relationship of the structure and function in biological system. Moreover, these hybrid devices could serve as an optical handle to investigate the properties of the structures themselves, such as how to control their optic-/electric- response for spectroelectrochemistry, which are widely used in (bio)analytical chemistry and interfacial science. In this field, the (electro)depositon of novel SRES-active (bio)structures directly on electrodes become a highlight and exciting goal, which produces inexpensive, controllable, reproducible, reusable biointerfaces by tailoring the morphology of the deposits, biomolecules, or combined both together. This chapter reviewed recent advances in the construction and applications of the biomaterials, mainly contributed by the authors' laboratory in the latest five years. The authors roughly describe several typical optobioelectronic devices, which combine biotechnology with optical (eg. surface enhanced resonance Raman scattering, SERRS) and electrochemical techniques, while the rapid progress of the nanotechnology provides extra opportunities for investigations in this field. Firstly, the authors discussed the recent development of the SERS-active nanostructures. These nanostructures are functionalized with biocompatible coatings focused on different mecaptoalkanes for efficient immobilization of the target protein. Then, these biosensing devices were applied to study the conformation change, redox behavior and biocatalysis of the immobilized proteins. Besides, the electron transfer mechanism of the protein with the conventional electrode was also discussed, according to the experiments and interfacial bioelectrochemistry theory. Furthermore, the comparable results, obtained from spectro-electrochemistry experiments, are explored and discussed in some detail. Finally, some significant advances combined electrochemistry with surface-sensitive SERS for the nanostructured biointerfaces are also outlined.

Chapter 3 - The authors present high-resolution electron energy loss spectroscopy (HREELS) measurements on surface plasmon dispersion in systems exhibiting quantum well states, i.e. Na/Cu(111), Ag/Cu(111), and Ag/Ni(111). The authors' results demonstrate that the dominant coefficient of surface plasmon dispersion for thin and layer-by-layer Ag films presenting quantum well states is quadratic even at small $q_{||}$, in contrast with previous measurements on Ag semi-infinite media and Ag thin films deposited on Si(111). The authors suggest that this behavior is due to screening effects enhanced by the presence of quantum well states shifting the position of the centroid of the induced charge less inside the geometrical surface compared to Ag surfaces and Ag/Si(111). For ultrathin Ag films, i.e. two layers, the dispersion was found to be not positive, as theoretically predicted. Annealing caused an enhancement of the free-electron character of the quantum well states, thus inducing a negative linear term of the dispersion curve of the surface plasmon. Moreover, the authors report the first experimental evidence of chemical interface damping in thin films for K/Ag/Ni(111). As regards Na/Cu(111), the authors found a different dispersion curve compared to thick Na films, thus confirming the enhanced screening by Na quantum well states. Results reported here should shed light on the influence of quantum well states on dynamical screening phenomena in thin films.

Chapter 4 - A novel photolithographic technique using a periodic hexagonal close packed silver nanoparticle 2D array photo mask has been demonstrated to transfer a nano-pattern into a photoresist using I line proximity photolithography. A 30 nm hexagonal close packed silver nanoparticle 2D array pattern with a 50 nm period has been successfully transferred into S1813 photoresist using I-line exposure wavelength. The resultant feature sizes were 34 nm with a period of 46 nm, due to the surface plasmon resonance where the S1813 photoresist feature is approximately 11 times smaller than I-line exposure wavelength.

Chapter 5 - Metal nanoparticles possess localized surface plasmon resonances (LSPR), which leads to unique effects, such as multiphoton-induced luminescence, radiative quenching or enhancement and spaser. In particular, gold nanorods (GNRs) are potential candidates for examining all these LSPR effects. GNRs have longtudinal surface plasmon band in the near infra-red (NIR) region, where the absorption of water and biological tissue is minimized, making them useful in biological research.Moreover, this longitudinal plasmon band can be tuned by changing the size, shape of nanorods and the dielectric environment. In this chapter, the authors review their work on surface plasmon enhanced two-photo luminescence (TPL) and energy transfer effect. The application of GNRs in biomedical imaging and analytical studies has been examined, together with a survey on localized surface plasmon related fluorescence enhancement and recent development on spaser.

Chapter 6 - The development of new technologies capable of monitoring the delicate details of the dynamics of living cells, which remain outside the optical resolution, is vital for cell research. The infrared light introduces several unique advantages for cell studies. Its interaction with live tissue causes no photodamage and allows examining chemical information encoded in infrared absorption spectrum. This chapter presents newly developed infrared surface plasmon resonance (SPR) methodology. This chapter discusses a specific aspects related to studies of aqueous samples using infrared surface plasmon. The extreme sensitivity of the surface plasmon to refractive index changes together with extended probing depth in the infrared range in comparison to visible range allows studying living cell morphology with submicron precision. Analysis of the temporal variation of the SPR reflectivity allows reconstructing the dynamics of the cell shape and observing various stages in cell behavior. Using infrared SPR technique the authors have studied the dynamics of cellular processes and successfully monitored submicron changes in cell morphology.

Chapter 7 - A new class of nonradiative surface plasmon polaritons guided by the interface between two enantiomeric anisotropic chiral plasmonic media is presented. The waves are caused by bianisotropy and are absent in the case of enantiomorph bi-isotropic materials as well as in anisotropic structures without magnetoelectric coupling. The basic properties of the waves, in particular, the dispersion relations, localization conditions and propagation lengths differ dramatically from those for conventional surface plasmon polaritons propagating along the interface between different achiral media.

Chapter 8 - Raman spectroscopy technique was applied to study phonon-plasmon coupled modes in Si doped set of GaAs/AlAs superlattices (SLs), grown on (100), (311)A and (311)B substrates. Anisotropy in dispersion of phonon-plasmon modes with wave-vector normal to SL layers and in SL layers was observed for (100) SLs. Additional lateral anisotropy of the phonon-plasmon modes was observed for SLs grown on nano-corrugated

(311)A surface. The observed anisotropy is supposed to be result of charge carrier mass anisotropy appearing due to structural anisotropy.

In: Plasmons: Structure, Properties and Applications
Editors: Anton E. Turunen and Johannes O. Niemi
ISBN: 978-1-62100-060-0

Chapter 1

Plasmonic Biosensors for Sensitive Measurements in Complex Biological Matrices

John S. Mitchell
The New Zealand Institute for Plant and Food Research Ltd, Hamilton, New Zealand

Abstract

Biomolecules of medical or physiological significance are often present only in very low concentrations in biological fluids such as blood, urine or saliva, making them difficult to measure. Such complex matrices pose particular challenges, such as fouling of sensor surfaces, obstruction of specific binding interactions, such as that between an antibody and its corresponding antigen, and production of non-specific binding responses in optical biosensors. Overcoming such challenges is central to the future usefulness of optical biosensor technology in fields such as medical diagnosis, physiological monitoring and the assessment of human health and well-being, particularly in non-invasive techniques where analyte concentrations are typically very low. Plasmons have been very widely utilized in transduction of biochemical binding events on sensing surfaces in the form of surface plasmon resonance (SPR) biosensors. Very sensitive assays can result when such a transduction is coupled with immunoassays, immunomimetic systems such as molecularly imprinted polymers, or complementary DNA binding interactions. These assays are capable of measurement at physiologically relevant concentrations in the blood and sometimes even in non-invasively obtained fluids such as saliva.

This chapter will examine the cutting-edge approaches that have been taken in adapting plasmonic biosensors to function in the presence of complex biological matrices. Most of the work in this area has focused on measurement in human blood but there will also be a discussion of emerging techniques for non-invasive measurement using fluids such as saliva and urine. The physicochemical challenges posed by these matrices will be outlined and related to the sophisticated chemical architectures that have been developed to minimize non-specific binding and ideally to help to ensure optimal sensor signal sensitivity. The functioning of the strategies applied will be related to surface chemistry and plasmonic theory. Included in this analysis are the varied physical separation and purification technologies used, where these are integrated with the sensor in an on-line approach. The analytes examined will include proteins, DNA and small molecules. Whilst the predominant emphasis will be on immunoassays, alternative binding interactions such as DNA binding and immunomimetic approaches will also be examined.

INTRODUCTION

Surface plasmon resonance (SPR) is a quantum optoelectronic phenomenon that can occur when a photon of light interacts with the electrons of a noble metal surface. When an incident photon couples with the electrons, its energy can be transferred to them so that they move as a single electrical entity known as a plasmon. This movement creates a surface electromagnetic wave that propagates parallel to the interface between the noble metal and a dielectric in contact with it. This generates an electrical field that extends out from the metal surface. Its intensity decays exponentially, with a typical significant field strength range of approximately 300 nm. When the metal surface is fixed to a single plane, SPR occurs at a well defined angle of incidence of the photon to the surface, known as the resonance angle (θ_r). The resonance angle is the angle required for the momentum of the photon to match that of the plasmon. This angle is a property of the composition of the surface within the electric field, so any mass that binds to or dissociates from the surface will perturb the resonance angle. Only small changes in bound mass are needed to produce large changes in the resonance angle, which is effectively measuring the refractive index of the surface. The most common experimental set-up used in plasmonics experiments is the so-called Kretschmann configuration [1], where a thin film of the noble metal (typically gold) is coated onto a glass slide. This is illuminated from the back-side with *p*-polarized light which penetrates the metallic film and generates the evanescent wave. Whilst some SPR instruments scan through a range of angles to determine the resonance angle, it is quite common for instruments to have fixed optics where the intensity of the light is monitored to determine a signal response. This is typically measured in resonance units (RU), where 1 RU = 1×10^{-6} refractive index units. A typical SPR sensor set-up is illustrated schematically in Figure 1.

The very high sensitivity of SPR to changes in bound mass on the metal surface makes it an excellent transduction technique for the detection of binding events between biomolecules, particularly where the biomolecule targets have a high mass themselves. It is possible to apply a wide range of different chemical treatments and surface layers to the gold surfaces to anchor one partner in the binding interaction to the sensing surface whilst flowing the other binding partner, in solution, over the surface and observing the change in signal response as a function of time. This generates a plot known as a sensorgram, which is acquired in real-time as the binding takes place. This has the benefit of being able to compute key kinetic parameters for the interaction and has led to commercial SPR brands becoming well-established in biomolecular interaction analysis. BIAcore^{TM} is perhaps the most widely known of these but there are also other companies offering simpler and more affordable options [3]. The types of interactions of particular interest have been antibody/antigen binding, DNA / complementary DNA interactions, host / guest interactions and more generalized protein / protein binding. SPR has also been coupled with biomimetic substrates such as molecularly imprinted polymers, in an attempt to replace antibodies [4].

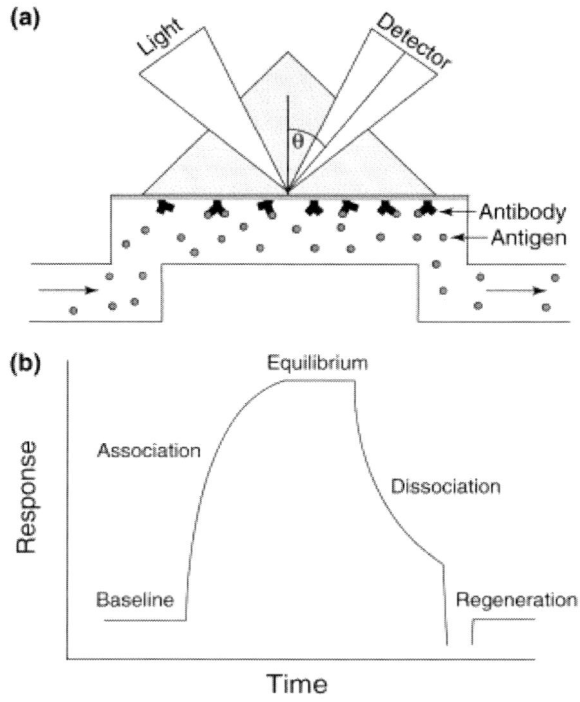

Figure 1. A schematic of a typical SPR set-up showing sample fluid flow over an antibody-immobilized surface and a corresponding sensorgram of binding response vs. time showing association, equilibrium, dissociation and regeneration stages [2]. Reprinted from Pharmaceutical Science and Technology Today, 3(9), Myszka, D.G., Rich, R.L., Implementing surface plasmon resonance biosensors in drug discovery, 310-317, Copyright 2000, with permission from Elsevier.

As well as examining biomolecular interactions, it is easy to see that SPR can be used as a transduction tool in assays, e.g. designing immunoassays on SPR surfaces or detecting complementary DNA strands in human samples. A very broad range of such assays have been developed, targeting a wide variety of different analytes, both large molecules such as proteins, and small molecules such as steroid hormones, drugs, environmental pollutants, toxins and vitamins [5]. In all these cases, at least one binding partner must be immobilized onto the sensor surface, preferably in a way that optimizes its presentation to its other binding partner, to ensure maximum signal response sensitivity. Based on these requirements, a number of different assay formats have been developed to achieve maximum sensitivity, low limits of detection (LOD) and stable sensor surface construction. It is also typically desirable to have a sensor surface that can be re-used for multiple concentration determinations, and so being able to regenerate the surface for re-use without damage to the immobilized components is important.

For DNA interactions, it is comparatively simple to immobilize a complementary strand of DNA to the sensor surface, but in the case of immunoassays a broader range of formats is used. When the analyte in question is a protein, then it is possible to construct a sandwich immunoassay format, whereby an antibody that binds a particular epitope of the antigen is immobilized on the metal surface. The protein analyte from the sample binds to the immobilized antibody, causing an increase in signal response and then, if added signal sensitivity is required, a secondary antibody can bind another epitope on the antigen in a so-called sandwich format, Figure 2A. If the analyte is a small molecule, then two main approaches are possible. An antibody can be immobilized on the sensor surface and the small molecule analyte in the sample can then be bound to the antibody, either sequentially or in competition with a labeled version of the analyte consisting of a large mass label attached to the analyte via an intermediate linker. Where the unlabeled analyte is exposed to the surface first, un-bound sites are then subsequently occupied by the labeled analyte conjugate, Figure 2B. An alternative approach is to immobilize the analyte via a suitable linker compound and then expose it to the antibody along with the sample containing free analyte. The free analyte will bind to some of the antibody molecules whilst leaving others to bind to the immobilized sensor surface, Figure 2C. This approach is useful in that immobilized small molecule antigen is usually more stable than immobilized proteins and so the surfaces are typically more robust towards chemical regeneration. Regeneration is achieved by exposing the sensor surface to a mixture or "cocktail" of reagents that can include extremes of pH and use of chaotropic reagents to remove the bound material from the immobilized surface.

Not all SPR biosensors use planar gold surfaces. There has been much interest in producing nano-islands or immobilized nanoparticles of gold or silver on a background substrate such as glass to make use of localized surface plasmon resonance (LSPR). Such approaches are favored as they have a steeper field strength gradient arising from the localization of the surface plasmons and so can be significantly more sensitive to mass changes on their surfaces than with a simple planar noble metal surface [6]. This approach also allows arrays to be produced where each island can have specially tailored surface chemistries and carefully selected immobilized analytes to enable simultaneous analysis of a number of different analytes by using imaging SPR (SPRI) to detect SPR responses in 2-D, typically with a CCD camera.

To be of functional use in the analysis of real samples, an SPR biosensor must be able to measure the analyte of interest in the midst of a complex background sample matrix. In biological samples such as blood, saliva, urine and synovial fluid, there will be high concentrations of both salts and high mass molecules such as proteins and mucins and in some cases even membrane fragments and cellular detritus may be drawn through the flow cell of the biosensor and over the sensor surface. High mass material is a particular challenge because if it non-specifically binds to the sensor surface it will produce significant signal not associated with binding of the analyte. Furthermore, such species may also sterically obstruct the binding interaction itself. Together, these factors can lead to highly variable sensor responses and a breakdown in the relationship between analyte concentration and signal response. They may also lead to a high level of binding signal even in the samples that should give low signal. This causes the active signal range to become compressed and results in a loss of signal sensitivity. If material continually binds non-specifically to the surface and is also hard to regenerate then it may eventually fully coat the sensor surface and prevent specific analyte binding, resulting in a total loss of surface functionality.

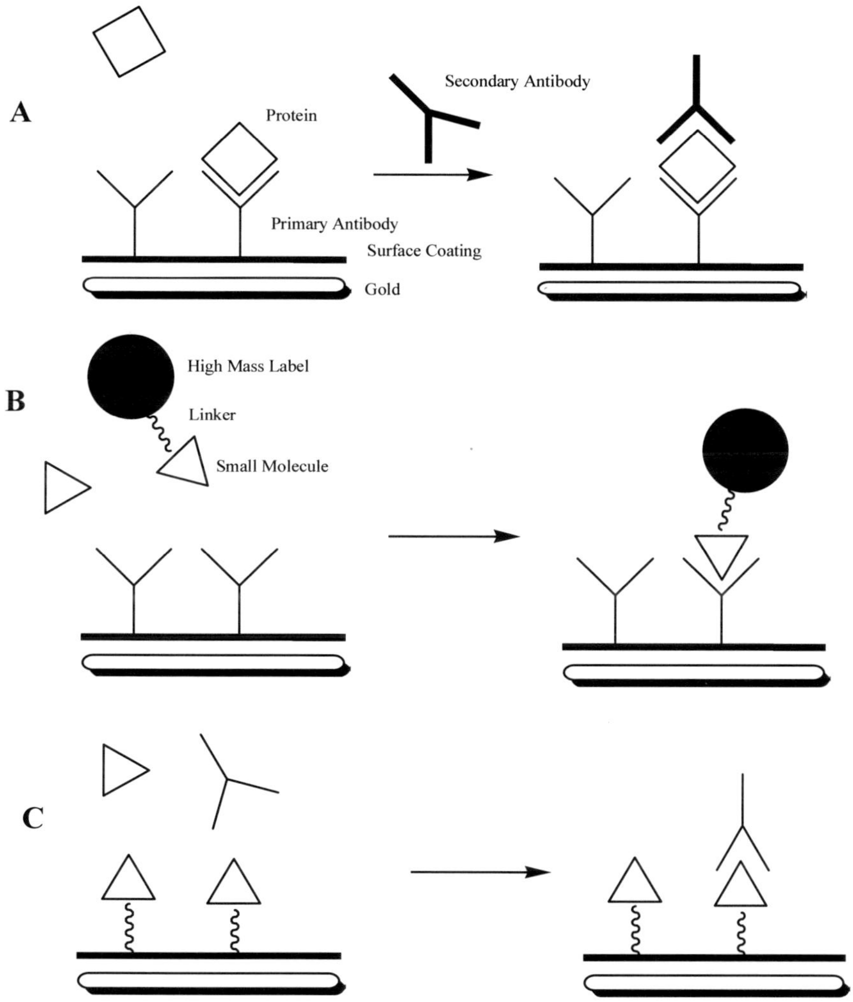

Figure 2. Schematic showing SPR detection formats, a. Sandwich assay of proteins, b. Antibody immobilized assay of small molecules, c. Antigen-immobilized assay of small molecules.

There is a pressing need to develop and implement new technologies to analyze compounds present only in very low concentrations in complex biological matrices. The comparatively high sensitivity of SPR techniques makes them a prime candidate as part of the next wave of devices to detect molecules for biomedical diagnostics, environmental monitoring and physiological investigations, but to work effectively they must overcome the major challenges posed by non-specific binding (NSB) to sensor surfaces of substances from complex biological media. This can be achieved by one or any combination of:

1) Physically removing the interfering compounds before passing over the sensor surface.
2) Designing surfaces that repel interfering substances in the background matrix and thus prevent NSB.

3) Adding reagents to the assay solutions to prevent or minimize non-specific interactions.
4) In cases where NSB is found to be predictable, mathematically subtracting this from the sensor response, typically using a reference channel.

Perhaps the most common approach to combating NSB, however, is to dilute the sample sufficiently to minimize interferences. This is of course only possible in cases where the analyte is of sufficiently high concentration and/or where the biosensor has been constructed with very low LOD; and often the dilution factors involved are very high [7]. Such an approach is therefore strictly limited in its application and represents no real additional challenge to the analyst beyond optimizing signal sensitivity; as such it will not be dealt with in any great detail in this chapter, other than to say that it is an option when sample and sensor parameters allow. In recent years there has been an increasing interest in chemical and physical approaches to addressing matrix interferences as a key aspect of practical plasmonic biosensing where low analyte concentrations do not permit extensive sample dilution.

In this chapter, I will focus on innovative approaches that have been taken to adapting plasmonic biosensors to function in complex biological matrices, regardless of the analyte being measured. It will look at the fundamentals of how surfaces and assay designs can be engineered to minimize this problem and enable effective assays. Sensor surface architectures based on self assembled monolayers (SAMs), intermediate linker compounds, DNA and polymer coatings will be examined and their performance discussed. The discussion will then move on to specific matrices such as blood, saliva and other fluids and the challenges posed in each instance. The chapter will conclude with a summation of the current state of the art and some comments on the future direction of plasmonic sensor design for real world sample analysis.

Fundamental Surface Design Concepts

Self assembled monolayers (SAMs) are chemical layers that form on a solid surface from molecules in solution by using some particular chemical functionality for their attachment. For gold surfaces, this is typically an organic straight chain molecule with a thiol (SH) terminus to allow coordinate binding to the gold. The chains can then contain other functional groups that allow immobilization of biomolecules. In the case of SAMs in SPR biosensors, it is quite common for researchers to apply a mixed SAM, i.e. one where alkane-thiol chains terminating in an appropriate functional group for immobilization are mixed in a fixed ratio with blocking thiols that contain non-reactive terminals [8]. By covering the surface with thiols it is hoped that this will sterically and chemically prevent molecules from adsorbing onto the gold surface, whilst a higher proportion of blocking to active thiol will provide the necessary space to allow biomolecules to access the immobilized binding partner and thus optimize signal sensitivity. This approach has been applied to the immunoassay of trinitrotoluene (TNT) with immobilized TNT analogues examined for optimizing LOD [8], Figure 3. SAM-based approaches have also been applied to the detection of bacteria in a multi-channel system using ethylene glycol-containing SAM chains to minimize NSB. However, in the case of this cellular matrix and in fruit juices, the approach was only partially effective and required a specific secondary antibody (one that binds to the primary antibody) to achieve adequate specific signal [9].

SAMs can also be applied to gold nanoparticles, and with an appropriate amine functionalization, they can be used as linkers to attach the nanoparticles to a surface and block the surfaces of the nanoparticles against NSB [10]. The length of the SAM chain, the degree of gold surface roughness and the length of time used for incubating SAMs with the surface can all have significant effects on the level of NSB and these parameters have been optimized for NSB of fibrinogen and lysozyme [11]. Passivation of the gold surface using simple disulfide-biotin coating has been shown to be effective in forming a SAM against NSB in assays of large molecules such as immunoglobulin G (IgG) [12]. Mixed oligoethylene glycol (OEG) SAMs have seen further application in reduction of NSB for sensitive detection of prostate specific antigen-alpha(1)-antichymotrypsin (PSA-ACT) complex in human serum with the ratio of active and blocking chains carefully optimized [13].

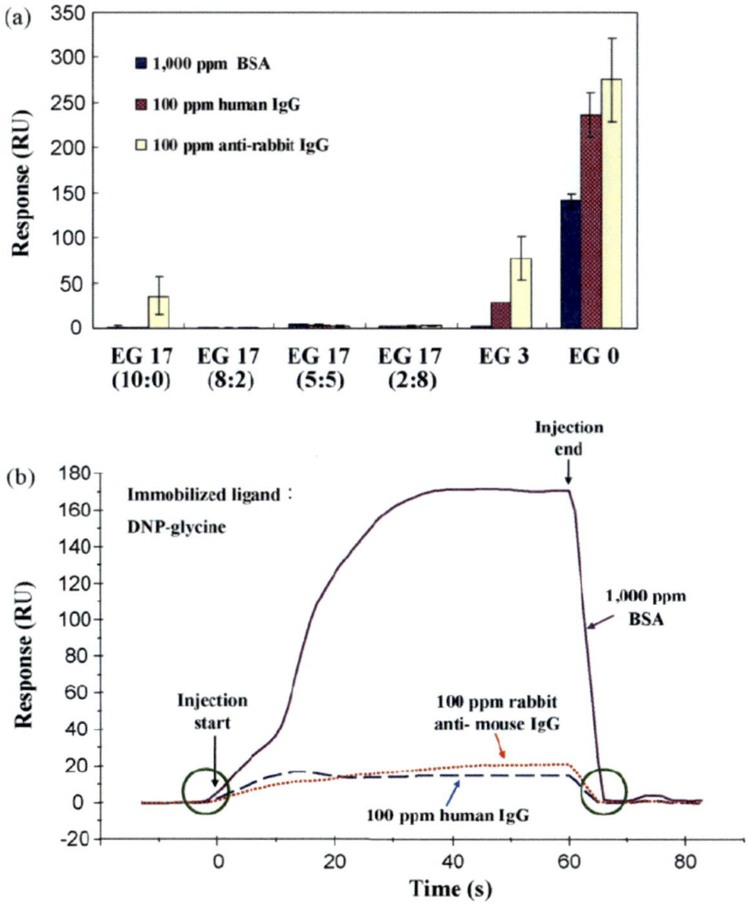

Figure 3. A. NSB from three proteins over mixed thiol SAM surfaces of different amine: blocking alcohol ratios (EG 17), an all carboxylic acid SAM (EG 3) and an all amine SAM (EG 0). B. Sensorgrams of NSB over an EG 17 chip (8:2 ratio) [8]. Reprinted from Biosensors and Bioelectronics, 24(2), Mizuta, Y.; Onodera, T.; Singh, P.; Matsumoto, K.; Miura, N.; Toko, K. Development of an oligo(ethylene glycol)-based SPR immunosensor for TNT detection, 191-197, Copyright 2008, with permission from Elsevier.

Dextran hydrogels are very commonly used as SPR immobilization layers and some reports have indicated that the inclusion of carboxymethyl (CM) dextran sodium salt in the diluent buffer can significantly reduce NSB in media such as sheep serum where in some instances it can reduce NSB from around 1000 RU to near 0 RU [14], Figure 4. Thin film deposition onto sensor surfaces has seen some success, with a protein resistant ethylene glycol-based polymer being deposited using an RF plasma to resist bovine serum albumin (BSA) fouling in planar SPR [15]. An interesting alternative approach has been to attach a protein sourced from mussels, which is notorious for its fouling properties, to the sensor surface using a polyethylene glycol (PEG) chain and using this to block the surface towards further fouling [16].

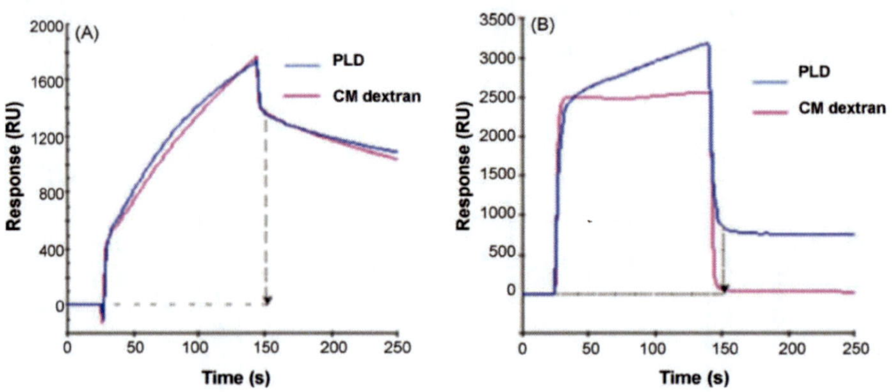

Figure 4. Sensorgrams for a positive serum control for determination of antibodies to phospolipase D (PLD) on functionalized sensor channels and dextran control channels using: A. No CM dextran sodium salt additive, B. With CM dextran sodium salt additive [14]. Reprinted from Analytica Chimica Acta, 651(1), Stapleton, S.; Bradshaw, B.; O'Kennedy, R. Development of a surface plasmon resonance-based assay for the detection of *Corynebacterium pseudotuberculosis* infection in sheep, 98-104, Copyright 2009, with permission from Elsevier.

Careful consideration of the possibility of NSB to antibody-immobilized surfaces is also crucial when developing multiplexed assays that detect multiple antigens simultaneously, particularly where the antibodies are applied as a single mixture [17]. In such cases, NSB could be minimized by selecting antibodies that exhibit low cross-reactivity, particularly to the other analytes of interest in the sample. When using SPRI, it is possible to introduce a 2-D variation in immunoreactivity across the sensor surfaces and thus create a gradient of response when a positive sample is applied, which is not the case with non-specifically bound material [18]. The use of surface plasmons with differing field distributions is an interesting approach to spectral discrimination of specific binding interactions from NSB in a multi-channel format [19]. In another example of multiplexed assays, human chorionic gonadotropin (hCG) and human albumin (hA) were detected without cross-reaction from NSB in human urine regardless of whether labels were used on just one or both analytes [20].

Inclusion of a reference to subtract NSB binding responses is often very useful. Molecularly imprinted polymers (MIPs) can be integrated with SPR transducers and their NSB properties can be quite high, so some form of correction by inclusion of a blank polymer as a reference is necessary and has been employed in the detection of viruses [21]. In the case of SPRI, careful referencing of the SPR responses in the flow chamber is very helpful in compensating for NSB effects and a Y-shaped microfluidic channel has been proposed to help do this [22], Figure 5. Using a microchamber array 2-D SPR set-up, it is possible to simultaneously observe both specific and non-specific binding using a bovine serum albumin (BSA) antigen [23]. Referencing has also been considered when using hydrogels and experiencing non-specific responses from solvent effects [24].

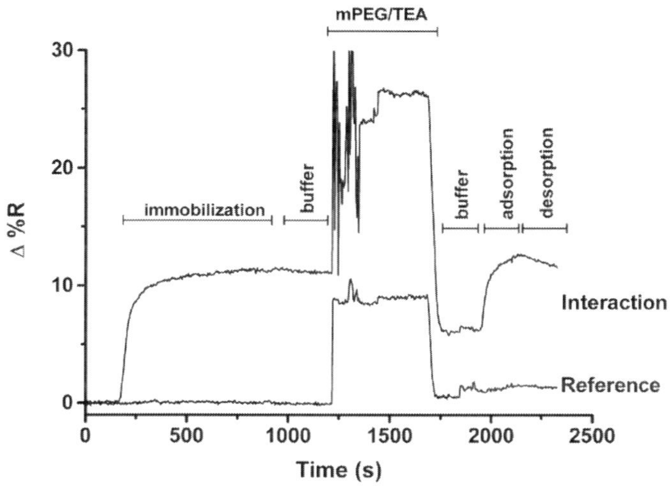

Figure 5. Sensorgram traces for the SPRI interaction region and a reference region showing immobilization processes and adsorption and desorption of an SPRI system for the interaction of *Datura stramonium* Agglutinin using a Y-shaped microfluidic channel [22]. Reprinted from Microchemical Journal, 93(1), Grasso, G.; D'Agata, R.; Zanoli, L.; Spoto, G. Microfluidic networks for surface plasmon resonance imaging real-time kinetics experiments, 82-86, Copyright 2009, with permission from Elsevier.

It has been found that keeping reference and sample channel immobilization densities similar to maintain similar charge states in the hydrogel and using conventional reference flow cell subtraction provides better control for solvent effects than attempts to make mathematical calibration corrections [24]. Referencing against NSB is also favored when the SPR surface is in the form of a spot [25] rather than a continuous planar surface.

The type of SPR transduction set-up used can also affect the levels of NSB, with a homemade grating SPR system reportedly giving lower NSB than a reference prism-coupled system when immobilizing anti-C-reactive protein fragments directly to the gold surface [26]. In switching between brands of SPR sensor, one must be conscious of the need to correct for potential NSB effects. This can be helped by use of an appropriate reference flow channel in semi-portable Spreeta^{TM} systems, which can also help to correct for temperature and bulk refractive index changes [27]. A sophisticated but expensive approach to overcoming NSB is to hyphenate the SPR technique with matrix assisted laser desorption ionization mass spectrometry with time-of-flight detection (MALDI-TOF), whereby species bound on the SPR surface are subsequently subjected to mass spectrometric interrogation that can quantize the specifically bound analyte [28]. A similar approach has been applied to analysis of staphylococcal enterotoxin B in milk and mushroom samples [29].

Polymer-based approaches are widely used not only for immobilization but also for resistance to NSB. Early work in this field compared the use of direct immobilization techniques onto glass-type surfaces to those using intermediate immobilization layers such as chitosan, dextran and other hydrophilic polymers and has found that NSB could be reduced by 5–50% by use of polymer layers whilst specific antibody binding responses could be greatly improved [30], Figure 6.

Non-ionic hydrophilic polymer intercalation between immobilized antibody fragments has been shown to be effective in repelling species from binding onto the gold surface [31]. Surface initiated atom transfer radical polymerization (SI-ATRP) of poly(oligo(ethylene glycol) methacrylate) can produce a thin film layer that has resistance to biofouling but it can be comparatively difficult to achieve immobilization of biomolecules on them [32]. In some cases it is desirable to cross-link a hydrogel to another polymer to form an optimized polymer coating with low NSB and this approach has been adopted using carboxymethylcellulose and polyethyleneimine for quartz crystal microbalances (QCM) [33]. It may be extended to SPR. In all SPR formats, including LSPR, it is necessary to ensure that potentially reactive groups such as carboxylic acids and active esters on the sensor coatings are passivated before running the assay, to minimize covalent and charge-based NSB [34]. NSB can also be reduced by coating a protein repellent polymer between immobilized Fab' fragments on gold slides [35].

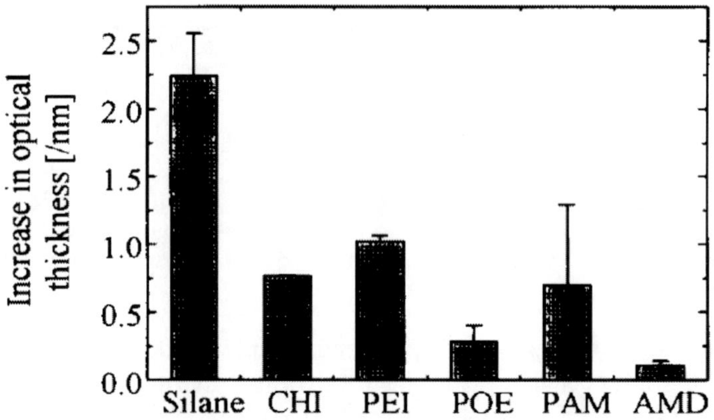

Figure 6. Levels of NSB for different surface treatments in SPR. CHI: chitosan, PEI: poly(ethyleneimine), POE: α,ω-diamino poly-(oxyethylene), PAM: poly(acrylamide-co-acrylic acid), AMD: aminodextran [30]. Reprinted from Biosensors and Bioelectronics, 11(6-7), Piehler, J.; Brecht, A.; Geckeler, K.E.; Gauglitz, G. Surface modification for direct immunoprobes, 579-590, Copyright 1996, with permission from Elsevier.

Linkers and intermediate immobilization layers can play a major role in surface functionalization. In some cases, immobilization to planar and nanoparticle gold surfaces can be achieved via special thiol linkers that contain two sulfur atoms at one terminal, e.g. thioctic acid, and such approaches have been shown to enhance specific and suppress non-specific protein binding in some cases [36]. In the case of DNA quadruplex folding, NSB can be a major issue that can affect data analysis if not handled carefully [37]. Modification of the immobilization support for antibodies in quantitation of myoglobin and cardiac Troponin I in human serum has proven effective in increasing sensor stability 5-fold and the study included testing of short chain thiols, polysaccharides and biocompatible polymers [38]. Intermediate layers composed of a trypsin inhibitor and glutaraldehyde have been advocated as a means of immobilizing proteins such as HSA on gold surfaces and have varying levels of NSB [39].

In some cases, silicon-based surfaces could be used in the sensor interface, such as in interferometry, and non-covalent immobilization of BSA can be effective in acting both as a linker and in reducing NSB [40]. Protein A can also serve both as a capture probe and a blocking agent to act against NSB of BSA to a porous silicon surface [41]. Similarly, a protein G layer used in immobilization of C-reactive protein (CRP) has shown no interfering signals from NSB [42]. Furthermore, glass modified for immobilization of antibodies via a photobiotin-avidin-biotin interaction has shown negligible NSB of photobiotin, ExtrAvdin and dye labeled anti-IgG [43]. Highly cross-linked Si-O-Si films with amine termini have been shown to be useful immobilization coatings of gold surfaces and have demonstrated 97% specificity when applied to measurement of avidin [44], Figure 7.

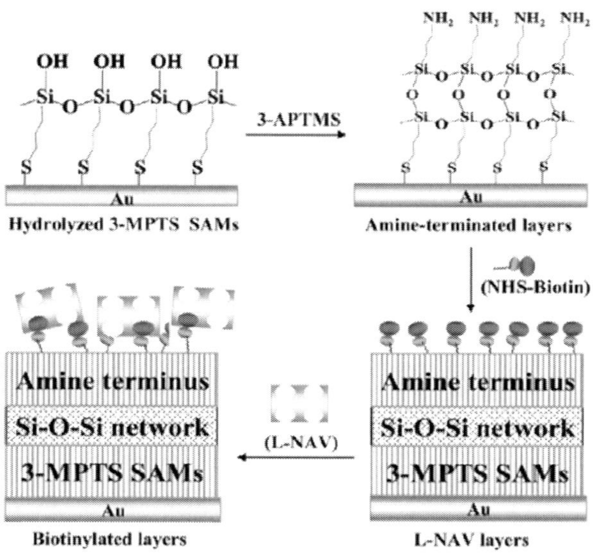

Figure 7. Highly cross-linked Si-O-Si film construction mounted on SAMs for detection of avidin [44]. Reprinted from Surface Science, 596(1-3), Liu, Z.; Amridis, M.D. FT-IRRAS quantitative analysis of specific avidin adsorption on biotinylated Au surfaces, 117-125, Copyright 2005, with permission from Elsevier.

NSB can also often be minimized by careful selection of the antibody used in immunoassay formats. Often it is necessary to base this selection on the effect of matrix interferences on assay parameters such as B_0 and IC_{50}, as in the case of urinary zilpaterol analysis [45]. It is worth noting that BSA is heavily implicated in NSB to gold surfaces, forming both reversible and irreversible films in a composite layer [46] and so is a major interfering substance examined in many of the assays mentioned in this chapter.

SELF ASSEMBLED MONOLAYERS (SAMS)

The application of SAMs to minimize NSB is widespread, and perhaps the most common approach is to use OEG-based SAMs, where their surface densities can be compared to their non-fouling properties for cases such as fibrinogen adsorption [47]. OEG chains are well-known for being non-immunogenic, biocompatible and hydrophilic [48], making them excellent candidates for SAM coatings both as blocking agents and as immobilization points. They are typically applied as mixed SAMs in an ethanolic / aqueous preparation. The SAM approach can also be applied to SPRI formats where it has been reported as a very successful method of suppressing NSB in analysis of a human glutamate decarboxylase isoform, to the point of not requiring reference subtraction [49]. SAMs can also be applied to gold nanoparticles immobilized on glass slides and used in LSPR, as in the case of 11-carbon chain thiols used to link to a beta secretase inhibitor where the mixed SAM significantly reduced NSB [50]. The mixed SAM approach has been shown to reduce non-specific hydrophobic interactions between enzyme and the gold surface when used for immobilization of esterase on gold nanoparticles [51]. Primary amines on proteins can be reacted with thioacetals at the termini of SAM chains to immobilize proteins such as PSA and in detection of *S. typhimurium* cells, and give minimal NSB to some potentially interfering biomolecules [52].

Quite low NSB responses to IgG were found when a gold-binding fusion protein was used to assemble a severe acute respiratory syndrome (SARS) surface antigen onto a gold surface, indicating that such gold-binding proteins can serve as both an anchoring agent and a blocking agent on the surface of the biosensor [53]. Use has also been made of modified SAMs such as *N*-hydroxysuccinimide terminated SAM on SPR biosensor surfaces [54]. 11-mercapto-1-undecanol (MUOH) and 11-mercaptoundecanoic-(8-biotinylamido-3,6-dioxaoctyl)amide (MBDA) can be used in a mixed SAM with a biotin functionality for attachment of ligands, and their mixing ratio can be optimized to minimize binding of streptavidin on the surface [55]. As a further modification to existing SAM technology, semi-fluorinated SAMs have shown greater protein repulsion than PEG-based SAMs and exert not only steric repulsion but also their lipophobic and hydrophobic fluorinated cylinder offers additional repulsive effects [56]. PEG modified with hydrizinehydrochloride and amine terminated has been effective in preventing NSB from proteins when used with the small molecule analyte TNT, allowing detection down to 8 pg/mL [57], Figure 8. Analogously, 2,4-dinitrotoluene can be immobilized using an OEG SAM and the resulting surface was found to be quite effective in preventing NSB from proteins [58]. Lipoate derivatives can be used to form imprints of small molecules on a sensor surface, though distinguishing between specific binding and NSB was not possible using liquid-phase simulations [59].

Optimization of the mixed SAM composition for minimizing NSB can also prove advantageous in improving assay signal, as was the case for PSA detection using a mixed SAM of 16-mercapto-1-hexadecanoic acid and MUOH at a 10% / 90% mix [60]. Liberal use of hydroxyl-terminated thiol SAM spacers was found to be effective in reducing NSB in the immobilization of biotinylated glutamic acid decarboxylase and concomitantly improving binding response [61]. The nature of the SAM chain definitely has a marked effect on the levels of NSB and this is reflected in the dominance of NSB response when methyl-terminated alkylthiolates are substituted for PEG-based chains in binding of streptavidin to biotinylated alkylthiolate SAMs [62]. Mixed SAMs, including biotin terminations, have been applied to construction of array surfaces for binding to streptavidin and are advocated for minimization of NSB in this format [63]. Alkyl thiolate mixed SAMs using a longer carboxylic acid terminated chain and a shorter ethylene glycol terminated chain have been studied for immobilization of 10 different proteins and were found to have a high resistance to NSB from these proteins across a broad range of isoelectric points and molecular weights [64].

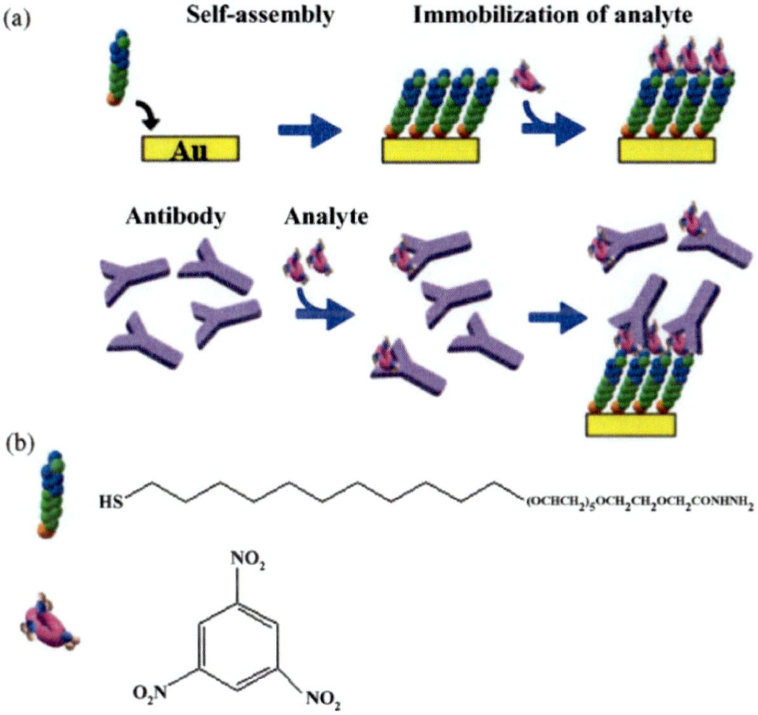

Figure 8. A schematic of the self-assembly of a SAM used in the detection of TNT and blocking of the sensor surface and the corresponding indirect competitive immunoassay format used [57]. Reprinted from Talanta, 72(2), Kawaguchi, T.; Shankaran, D.R.; Kim, S.J.; Gobi, K.V.; Matusmoto, K.; Toko, K.; Miura, N., Fabrication of a novel immunosensor using functionalized self-assembled monolayer for trace level detection of TNT by surface plasmon resonance, 554-560, Copyright 2007, with permission from Elsevier.

Linker Chemistry

The importance of considering the nature of intermediate linkers used in immobilization of sensor surfaces is illustrated by a study of an oligo(ethyleneglycol)-dithiocarboxylic acid derivative that was used as a heterobifunctional linker for immobilization of insulin in an SPR immunosensor and where the use of the OEG chain resulted in an assay with low reported protein NSB [65]. This is further reinforced by investigations of poly-(L-lysine) binding to SAMs of 11-mercaptoundecanoic acid (MUA), 1-undecanethiol and MUOH. Irreversible binding was seen in all three cases but the mechanism of interaction for MUA was electrostatic, involving the terminal carboxylic acid group, whilst the other two thiols did not achieve such a good surface coverage and thus allowed direct NSB to patches of bare gold surface [66].

By incorporating two zwitterionic poly(carboxybetaine) arms into a mussel biomimetic polymer, very low fouling could be achieved upon exposure to undiluted blood plasma and serum and such properties were maintained when antibody was immobilized to the polymer [67], Figure 9. When ganglioside-incorporated phospholipid vesicles were spread spontaneously onto an octane-thiol coated gold SPR sensor surface, they served as an immobilization matrix for detection of cholera toxin via chemiluminescence and exhibit low NSB from serum components despite the complexity of the matrix [68]. In the complete inverse of the usual desire to reduce NSB, lysozyme can be non-specifically bound to the bare gold surface to construct the sensor interface by directing the formation of a monolayer of silica nanoparticles which can encapsulate enzymes such as an organophosphate hydrolase [69].

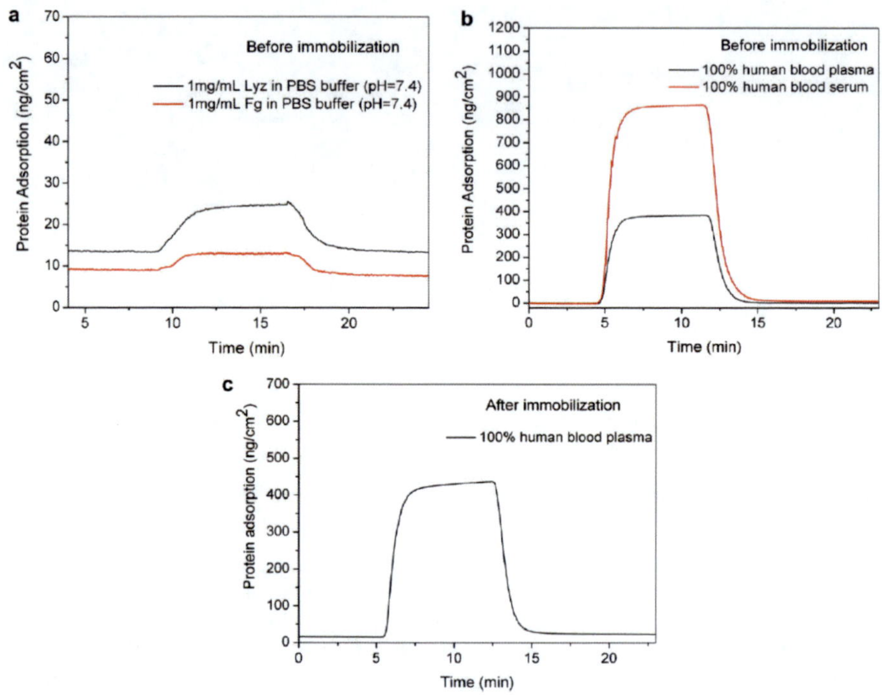

Figure 9. NSB responses over a polymer incorporating zwitterionic poly(carboxybetaine) arms, for: A. lysozyme and fibrinogen before antibody immobilization, B. human blood plasma and serum before immobilization, C. human blood plasma after immobilization; where the surface is used for detection of activated leukocyte cell adhesion molecule [67]. Reprinted from Biomaterials, 31(7), Gao, C.L.; Li, G.Z.; Xue, H.; Yang, W.; Zhang, F.B.; Jiang, S.Y. Functionalizable and ultra-low fouling zwiterionic surfaces via adhesive mussel mimetic linkages, 1486-1492, Copyright 2010, with permission from Elsevier.

The use of protein carriers as linkers to the sensor surface has also been explored. When protein conjugates of 2,4-dichlorophenoxyacetic acid (2,4-D) were coated onto a gold surface by elementary protein adsorption, the surfaces were reported to have low protein NSB where the carrier protein acts as a blocking agent [70]. Blocking agents can be injected over the sensor surface after immobilization but before the assay begins, such as in the case of a troponin T immunoassay where 10 mM glycine was used to block a CM dextran-based SPR sensor surface [71]. It is also possible to coat surfaces with a fluorocarbon thin-film that is hydrophobic and inert and this has been done on silicon nitride surfaces, but presumably could be useful for gold SPR surfaces also; in this case it is reported to prevent NSB [72]. By utilizing the high affinity interaction between Concanavalin A and glycoproteic enzyme, it is possible to immobilize acetylcholine-esterase in such a way that chemical cross-linking of proteins is avoided and NSB is constrained to around 3% of binding [73].

Amine-terminated polyamidoamine (G4-NH_2 PAMAM) dendrimers can be attached to gold surfaces via a SAM terminating in aldehyde groups constructed from 2-hydroxypentamethylene sulfide, and in examining biotin / avidin interactions, they show a relatively high binding specificity whilst retaining a high binding capacity [74]. A commercially available bifunctional coupling linker with a calixcrown moiety, ProLinker™ B, has been used to immobilize antibodies to detect a human parathyroid fragment that has been labeled with hexahisitidine-ubiquitin as expressed in *Escherichia coli* using an SPRI format [75]. No response was observed when un-induced cell lysate was passed over the surface, indicating good resistance to NSB [75]. It has also been previously demonstrated that when a site-directed immobilization is employed using hydrophilic SAM layers of either 4-aminothiophenol, mercaptopropionic acid, L-cysteine or thioctic acid, lower NSB can be obtained over silver surfaces than with simple random physical adsorption of IgG protein [76]. Immobilization of biotin as a chemisorbed monolayer on gold films has been reported as having no detectable NSB to the surface and was applied for binding of streptavidin [77].

Lipids with terminal linker groups can be used to immobilize Fab' fragments whereby the lipid is embedded in cholesterol and dipalmitoylphosphatidylcholine (DPPC) binary and ternary monolayers, and when DPPC is used, NSB to BSA is quite low [78], Figure 10. DNA can be anchored via a cholesterol tag to a lipid monolayer on a gold film as an alternative to conventional dextran-based immobilization where, as in the case of kinetics studies of high mobility group proteins, they can offer lower NSB [79]. This was done using phosphatidylcholine and demonstrated reliable kinetic binding interactions and low NSB [79].

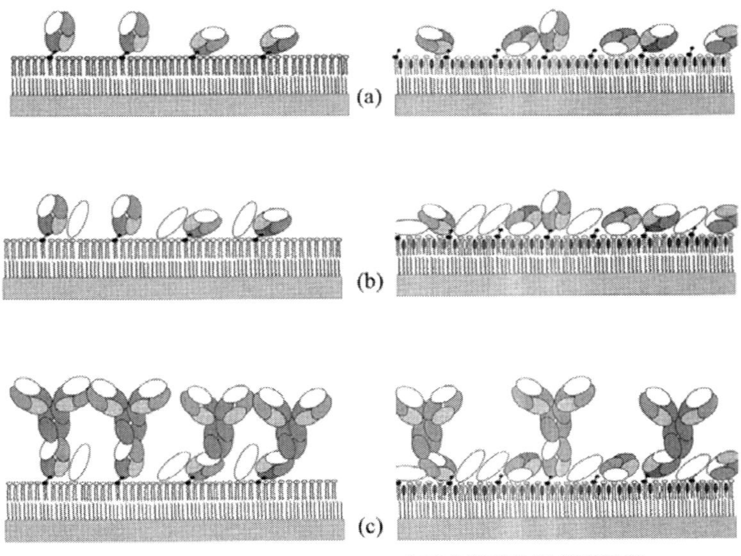

Figure 10. Schematic of surface construction on a lipid linker monolayer. A. Attachment of Fab' to the lipids, B. NSB of BSA, C. Attachment of human IgG. The left layer has no cholesterol included, whilst the right layer has cholesterol incorporated [78]. Reprinted from Biochimica et Biophysica Acta (BBA)-Biomembranes, 1421(1), Vikholm, I.; Viitala, T.; Albers, W.M.; Peltonen, J., Highly efficient immobilization of antibody fragments to functionalized lipid monolayers, 39-52, Copyright 1999, with permission from Elsevier.

POLYMERS

CM dextrans are a common default polymer used in SPR surfaces and demonstrate significant reductions in NSB proportionally related to polymer layer thickness and density [80], but it is worthwhile to consider alternatives. A reduction in NSB from serum proteins of up to 73% was observed when changing from CM dextran at 500 kDa to other polymers such as lower mass dextrans [81]. It was also possible to observe a 64% drop in BSA NSB whilst retaining most of the desired binding capacity when using an orthopyridyldisulfide-polyethylene glycol-*N*-hydroxysuccinimide (OPSS-PEG-NHS) polymer [81]. Such polymers can be attached to the surface via thiol linkers and compare well to more conventional dextran approaches [82]. Additionally, the influence of steric problems and re-binding issues in kinetics experiments using CM dextrans makes comparison between them and other coatings worthwhile [83].

Grafted high density polymer films have been developed based around zwitterionic polymer conjugates of carboxybetaine methacrylate attached to 3,4-dihydroxy-t-phenylalanine (DOPA) as a biomimetic moiety on silica surfaces [84]. Grafting conditions such as temperature, pH, salt concentration and buffer type could be optimized to minimize NSB and showed a high degree of resistance to NSB even in 100% human blood serum and plasma [84]. The grafting approach has also been applied to construct poly[2-(methacryloyloxy) ethyl phosphorylcholine] coatings to gold surfaces, where thicker layers demonstrated greater resistance to BSA adsorption [85]. Adsorption of grafted copolymers onto SAMs can be used to provide resistance to protein adsorption of BSA and fibrinogen, as in the case of poly(ethylene imine)-[g]poly(ethylene glycol) (PEI-PEG) and poly(L-lysine)-[g]poly(ethylene glycol) (PLL-PEG), where the later showed better protein resistance and where it was found that intermediate charge in the polymer combined with charge neutralization between the underlying surface and the polymer layer resulted in high resistance to adsorption of protein [86].

Some polymer thin films will spontaneously self assemble onto gold surfaces, such as a certain analogue of polyacrylamide, which is functionalized by coupling to 3-(methylthio)propylamine (MTP) [87]. In this example, resistance to protein NSB is a function of the level of MTP grafting in the polymer [87]. Multilayers of polyelectrolyte can be used to coat gold nanoparticles with a porous film and show significant alteration of protein NSB interactions compared with bare gold [88]. Graft copolymers composed of poly(ethylene glycol) and poly(L-lysine) in varying proportions and bearing nitrilotriacetic acid groups can be spontaneously assembled electrostatically from solution and demonstrate low NSB when exposed to human serum whilst retaining specific and reversible binding to a His-tagged green fluorescent protein target [89]. A graft polymer composed of poly(acrylic acid) (PAA) and a thiolated poly(vinyl)alcohol was reported to give a hydrophilic surface coating with low NSB and high binding capacity for IgG [90].

When a copolymer coating is applied that consists of both negatively charged carboxylic acid-containing monomers and positively charged monomer with quaternary amines, it is possible to tune the non-fouling properties according to the pH of the surrounding fluid and this can be applied to trap and later release species such as bacteria from the surface [91]. One common approach to combat NSB using polymers is to coat with block copolymers containing PEG chains. This approach is employed in a poly(ethylene oxide)/poly(propylene oxide)/poly(ethylene oxide) triblock copolymer where the ratios between the three components can be altered [92]. It was found that use of a Pluronic™ F108 coating showed high resistance to NSB from BSA, fibrinogen and complicated mixtures of proteins [92]. Polymer blend ratios have been shown to have an effect on NSB in the case of poly-L-lysine and polystyrenesulfonate polyion films. NSB of less than 10% that of the un-modified surface was obtained when the ratio of poly-L-lysine to polystyrenesulfonate was 4:6 [93] (Figure 11). These blends could be applied to direct immunoassay of a cytokine [94].

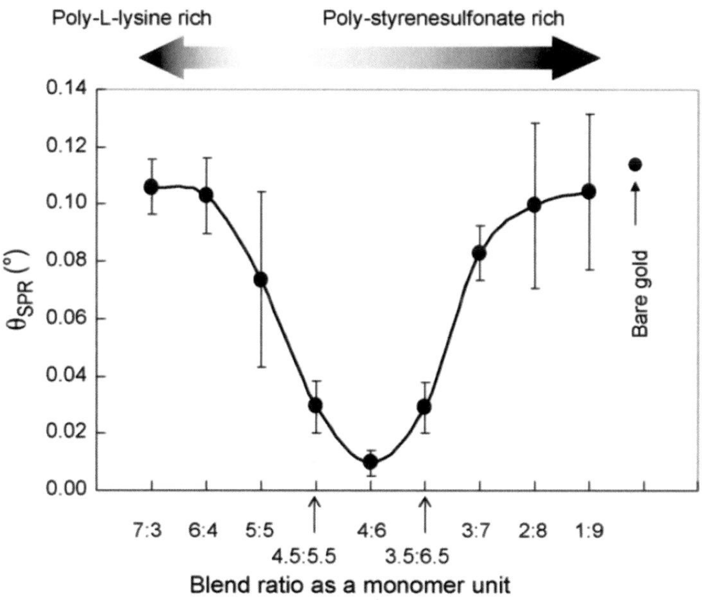

Figure 11. The effect of the blend ratio of polymerization on the levels of NSB seen on a gold SPR surface coated with poly-L-lysine and polystyrenesulfonate polyion films, clearly indicating the optimal blend ratio; the vertical axis is the resonance angle shift [93]. Reprinted from Sensors and Actuators B - Chemical, 130(1), Kurita, R.; Hirata, Y.; Yabuki, S.; Yokota, Y.; Kato, D.; Sato, Y.; Mizutani, F.; Niwa, O. Surface modification of thin polyion complex film for surface plasmon resonance immunosensor, 320-325, Copyright 2008, with permission from Elsevier.

High density PEG films have been examined by a variety of techniques in the case of 2 kDa PEG chains coupled to silica using glycidyloxipropyltrimethoxysilane (GOPTS) silanization [95]. No significant NSB was seen with 1 mg/mL ovalbumin exposure and only minimal (150 pg/mm^2) NSB from 10% calf serum, representing a 90% reduction compared with silanization alone [95].

One can also copolymerize monomers that terminate in thiol groups, such as in the case of 11-mercapto-*N,N,N*-trimethylammonium chloride and 11-mercaptoundecylsulfonic acid where polymerization occurs by surface-initiated atom transfer radical polymerization (SI-ATRP) and charge distribution is regulated by the rate of this polymerization [96]. Their resistance to protein fouling is dependent upon the ionic strength of the intra- and intermolecular interactions between the polymer and sample proteins [96]. The SI-ATRP approach can also be used to lay down poly(2-hydroxyethyl methacrylate) layers on silanized glass to resist NSB from *Salmonella typhimurium* bacteria, resulting for the most part from the hydrophilic properties of the interface [97]. ATRP can also be used where the initiator is immobilized by grafting onto the sensor surface and an even brush of sulfobetaine methacrylate is polymerized to form the thin film [98]. This has shown very low fouling and, in the case of fibrinogen, binding is less than 0.3 ng/cm^2 [98]. This produced much better results than an analogous hydroxyl terminated SAM-based approach [98]. The way in which analytes are immobilized onto sensor surfaces can have great effects on assay performance and NSB. For example, ATRP of 2-lactobionamidoethyl methacrylate (LAMA) containing pendant galactose residues as compared to a disulfide SAM with two galactose residues on the chain [99]. Binding of a lectin to the polymer was reversible, but for the SAM it showed strong NSB [99]. Radiolabeled proteins have been employed to examine NSB to *N*-isopropyl acrylamide-based polymer coatings and results suggest that temperature can play a significant role in NSB as the surfaces demonstrated low NSB at low protein concentrations and temperatures below the lower critical solution temperature but higher NSB above this temperature [100].

It is possible to employ more technically sophisticated approaches to layer design by using pulsed plasma polymerization to deposit thin films such as di(ethylene glycol) monovinyl ether to help resist BSA and fibrinogen NSB and allow DNA interaction signals to be discriminated [101], Figure 12. When using electrostatic interactions for cellular attachment to surfaces, such as those in SPRI sensors, it is necessary to ensure low fouling and this can be aided by use of a low fouling base polymer coat before attaching functionalized polymer coatings such as phenylazide-modified polymers [102]. Functionalized hydrogel networks can have anti-fouling properties, such as in the case of *N*-isopropyl-acrylamide-based polymers used for IgG immobilization [103].

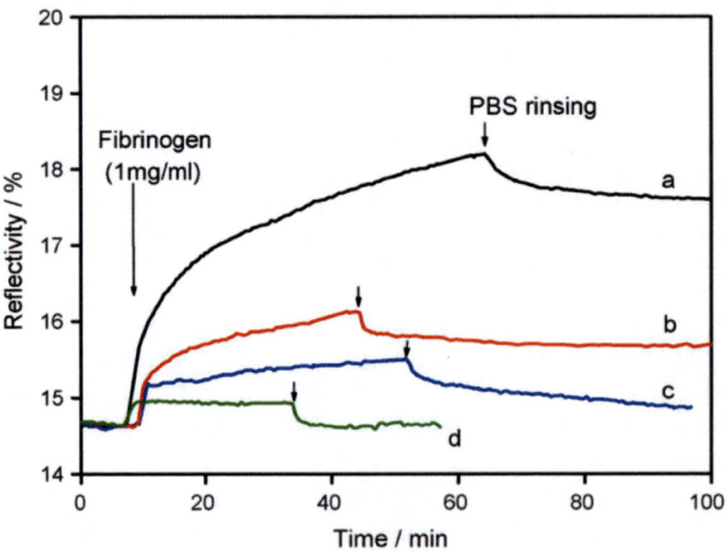

Figure 12. Sensorgrams of fibrinogen adsorption onto di(ethylene glycol) monovinyl ether thin film as a product of film thickness: a. 3.0 nm, b. 3.9 nm, c. 4.5 nm and d. 5.5 nm (all ± 0.2 nm) [101]. Reprinted from Biosensors and Bioelectronics, 25(2), Chu, L.Q.; Knoll, W.; Forch, R., Plasma polymerized non-fouling thin films for DNA immobilization, 519-522, Copyright 2009, with permission from Elsevier.

Technology applied to electrodes can also be useful in SPR. Electropolymerization of tyramine and pyrrole-1-propionic acid in a layer-by-layer construction produced no noticeable surface NSB in detection of dopamine [104]. Catechol groups can also be used to anchor biomimetic polymers to gold surfaces, either with or without functionalized intermediate SAMs and with zwitterionic moieties [105]. Such formats have been shown to resist bacterial adhesion and biofilm production and, furthermore, have shown resistance to protein NSB from fibrinogen, blood plasma and serum and lysozyme [105].

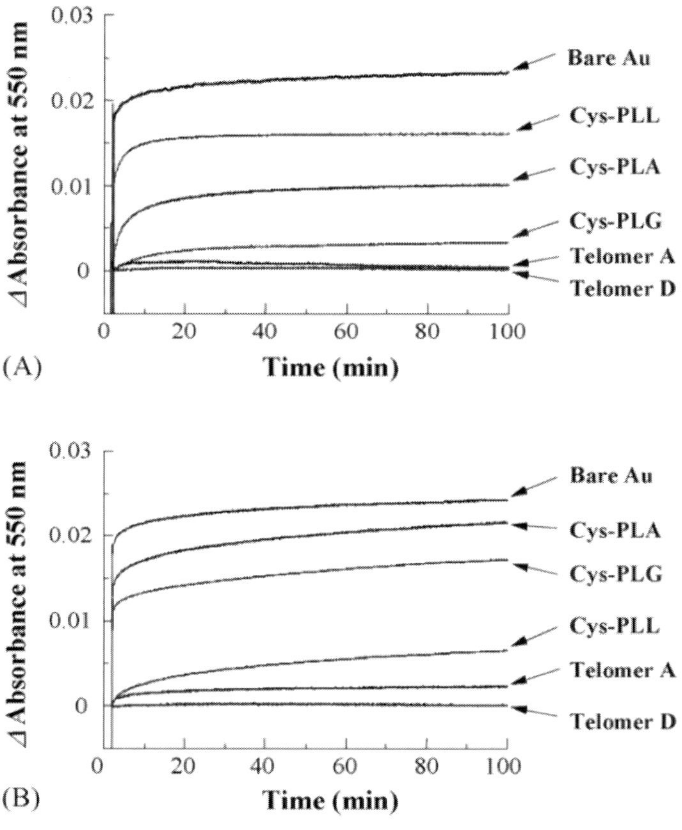

Figure 13. NSB responses for a selection of telomer coated surfaces to help optimize the coating, A. BSA 0.1 mg/mL, B. Lysozyme 0.1 mg/mL [112]. Reprinted from Colloids and Surfaces B: Biointerfaces, 56(1-2), Kitano, H.; Gemmei-Ide, M.; Anraku, Y.; Saruwatari, Y., Resistance of surface-confined telomers with pendant glucosylurea groups against non-specific adsorption of proteins, 188-196, Copyright 2007, with permission from Elsevier.

As a simple alternative, Tween-20 surfactant has been modified to incorporate thiol groups and has been applied as a thin film on a gold layer and showed strong resistance to protein NSB whilst allowing functionalization via hydroxyl groups [106]. A simple and comparatively inexpensive approach can be to use thioacetal linkages in 3-D polymer layers for immobilization, combined with thiol or thioether moieties for attachment to the gold surface, and these can result in low NSB of analytes [107]. Sometimes the polymers employed can be complex but the application technique simple, as in the case of solvent evaporation of thin films of poly[2-methacryloyloxyethyl phosphorylcholine (MPC)-co-n-butyl methacrylate (BMA)-co-2-lactobionami-doethyl methacrylate (LAMA)] (PMBL), where NSB from concanavalin A was reduced [108]. Similarly, a complicated phospholipid-based polymer, poly [2-methacryloyloxyethyl phosphorylcholine-co-n-butyl methacrylate-co-*p*-nitrophenyloxycarbonyl poly(ethylene glycol) methacrylate] (PMBN) coating can be used in QCM, and by extension possibly in SPR, to immobilize binding partners via NHS linkage and was reported to suppress NSB from proteins such as BSA and gamma-globulin on account of phosphorylcholine groups on the outer side of the polymer [109].

Single-stranded DNA can be attached along with polyacrylamide coatings either together or sequentially and the polymer was shown to lower NSB from DNA in samples [110]. In some cases, polymer can be used to in-fill the gaps between immobilized species, as in the case of immobilization of Fab and Fab' fragments with a *N*-[tris(hydroxy-methyl)methyl]acrylamide polymer used as a hydrophilic and non-ionic filler to reduce NSB [111]. Random telomers can be produced that have thiol groups and pendant D-glucosylurea groups using UV polymerization and can attach to planar gold surfaces or surfaces modified with gold colloids. In this case, the optimized polymer brush did not demonstrate significant NSB to BSA and egg white lysozyme and only slight NSB to a variety of sugar-binding proteins such as wheat germ agglutinin and concanavalin as opposed to certain pyranoside and glucopyranoside-containing coatings [112], Figure 13. When printing protein microarrays that can potentially be used in SPR applications, chemical treatments using tetraoctadecylammonium bromide can also prevent protein immobilization in undesirable areas by the application of a temporary polymer mask, which acts as a hydrophobic barrier on top of CM dextran. However, the barriers must be removed before assay as they could act as a source of NSB themselves [113].

DNA

The immobilization of DNA strands onto noble metal SPR biosensor surfaces is of increasing interest in developing high-throughput assays and array formats. Surface construction is also a major consideration in the determination of binding kinetics such as those for double stranded DNA fragments [114]. However, if DNA is simply immobilized via thiol modifications to their termini then significant NSB can result on the sensor surface [115]. The other techniques mentioned in this chapter can potentially be very helpful for DNA, but there are some reports that are more specific to DNA immobilization and so are discussed separately here.

Streptavidin can be used as a coating layer underneath oligonucleotide strands and has the effect of reducing NSB from nanoparticle binding to oligonucleotide strands as they are passed over the sensor surface [115]. In similar fashion, blocking polymers can be coated either simultaneously or sequentially with a single stranded DNA (ssDNA) strand with lipoate and hydrophilic polymers being tested [116]. The polymer layers can also impact on the binding between the complementary DNA strands, with best results coming from a mixture of lipoate and the ssDNA strands [116]. A coating of a graphite electrode that may also hold benefits for use in SPR is one based on ssDNA bound to a CM dextran layer. In this case, a mixture of sodium dodecyl sulfate (SDS) and milk powder is added to the hybridization solution to minimize NSB between the DNA strand and the glucose oxidase-avidin D conjugate that is used as a label in the interaction [117]. Mixed PEG chains have been used to produce SAMs that both block NSB from proteins through repulsion and also allow immobilization of ssDNA via a biotin / streptavidin interaction at the outer termini of the PEG chains in a fiber optic SPR format. In this format, negative control samples did not show any noteworthy NSB [118].

In some cases, non-complementary interactions can be distinguished from complementary or specific interactions by virtue of the format used in the assay.

This is the case with LSPR assays using colloidal nanoparticles of gold or silver, where the peak shift in the LSPR spectrum is specific to the interaction of interest, as only the complementary strands of DNA bear nanoparticles [119].

In other cases, complex and unwieldy chemistry can be used to attach DNA strands to the biosensor surface, such as the use of a supramolecular assembly of a thiol SAM, nickel-iminodiacetate functionalized phospholipid, modified cytochrome b5, heterobifunctional linker chain and then modified ssDNA, which is reported to have specific hybridization coupled with quite low NSB [120], Figure 14. In contrast, other reports have advocated using streptavidin monolayers that are close-packed on the gold surface and binding of simple biotinylated double stranded DNA and are reported to have low NSB to proteins in the sample [121].

Aptamers are oligonucleotides or peptides that bind specifically to a particular analyte and as such they can be quite useful as a binding partner in the construction of SPR assays. For such applications they are predominantly oligonucleotide chains. Aptamers can be immobilized onto gold nanoparticles and the resulting conjugate blocked with 6-mercapto-hexan-1-ol to minimize NSB in the detection of human IgE [122]. Another study has shown that for small molecules such as codeine, immobilization of the target directly to the surface, without a carrier protein, results in lower NSB when detection is based on aptamer binding. This is an important observation when the format has the aptamer free in solution rather than immobilized on the sensor surface [123].

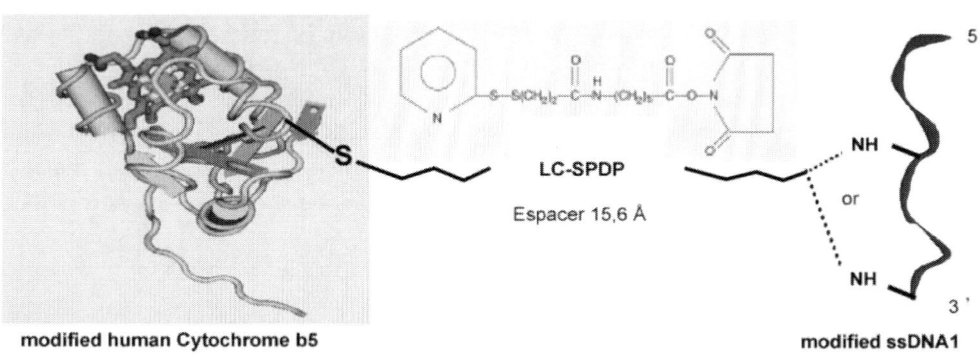

Figure 14. Use of a heterobifunctional linker chain to immobilize ssDNA on a sensor surface using a complex supramolecular assembly [120]. Reprinted from Biosensors and Bioelectronics, 20(8), Boireau, W.; Zeeh, J.C.; Puig, P.E.; Pompon, D., Unique supramolecular assembly of a redox protein with nucleic acids onto hybrid biolayer: towards a dynamic DNA chip, 1631-1637, Copyright 2005, with permission from Elsevier.

BLOOD

When analyzing chemicals in blood or blood products by SPR, some earlier work simply relied upon the standard running buffer composition (including EDTA and P-20 additives) to minimize NSB and checked that a related but different analyte did not give a significant binding response to the surface, as in the detection of α-Gal antibody interactions with immobilized linear blood group B saccharides [124]. Another earlier report suggested negligible NSB from erythrocytes when blood is passed over an immunoglobulin M (IgM)-immobilized surface in a commercial BIAcore™ instrument for analysis of blood type through cell / surface binding [125]. However, it is increasingly acknowledged that more thorough handling of NSB from whole blood and blood fractions is needed to ensure reliable assays when measuring in this complex matrix.

A detailed comparison between the blood plasma binding properties of diethylene glycol carboxylated SAMs, PEG-grafted SAMs and zitterionic methacrylate-based polymer brushes has found that high plasma binding onto unmodified SAMs could be significantly reduced by PEG grafting and NSB could be completely prevented by employing a poly(carboxybetaine methacrylate) coating polymer [126]. The PEG-grafted SAMs only gave NSB to proteins higher than 350 kDa [126]. The study also revealed that wettability and adsorption of individual plasma proteins is no indication of the performance of the surface for plasma as a whole [126].

Blood analysis using SPRI arrays is of growing interest as a high-throughput technique. Thiolated oligonucleotide strands have been immobilized by spotting on an SPR sensor surface and then filling the rest of the areas on the sensor surface with carboxylic acid terminated thiol SAMs, which then have BSA immobilized covalently to them to produce a low NSB surface for analysis of protein biomarkers in 10% blood plasma [127]. Binary patterned peptides can be used to produce monolayers that block NSB from proteins in crude bovine serum without the need to pre-treat samples. An optimized peptide was selected with a 3-mercaptopropionic acid head group for immobilization of antibodies [128]. Using this format for detection of matrix metalloproteinase 3 resulted in very low NSB from undiluted blank serum [128]. Zwitterionic polymers are valuable for use with measurements in blood plasma as they can produce functionalized coatings with quite low NSB. For example, poly(carboxybetaine acrylamide) (polyCBAA) has been used for immobilization of proteins and after 6 h exposure to blood plasma without dilution, gave less than 7 ng/cm^2 adsorbed protein [129], Figure 15. An earlier report of the same concept suggested NSB of less than 3 ng/cm^2 after exposure to blood plasma without dilution [130].

Thin graft CM dextran layers have been examined for their interactions with proteins in blood serum [131]. Careful surface examination revealed that such layers can be prepared with high density and flexible chains and that pinning density is a key factor in regulating surface coverage and amount of adsorbed material [131]. NSB occurring during measurement in bovine serum using a CM dextran chip has been examined using leptin as an example analyte and use of a modified buffer system with controlled NaCl, a high concentration of dissolved CM dextran (0.005%) and a pH of 9 gave a reduction in NSB of 86–87% on average [132]. An interesting approach for use with serum samples is the application of lipid vesicles to the sensing surface after immobilization of antibody. These vesicles can then fuse to form a membrane, and proteins from the serum non-specifically bind to the supported membrane, which is then removed by a non-ionic surfactant before measuring the antibody binding response [133].

When determining human growth hormone in human serum samples, alteration of the buffer composition has been shown to be effective in reducing NSB, with reductions of 88% reported when a high salt loading (550 mM), high Tween surfactant (0.1%) and an elevated pH of 8 are used instead of a more conventional buffer composition [134]. The buffer optimization approach has also been used when measuring 7-hydroxycoumarin in serum using a standard BIAcore biosensor and protein conjugate immobilization [135]. Detection of IA-2 autoantibodies in human serum has benefited from careful optimization of the ratio of mixed SAMs used on the sensor surface and the introduction of PEG chains into the SAM layers. An anionic surfactant additive can be used in the human serum samples and managed to contain the NSB from the matrix whilst maintaining the viscoelastic properties of the sensor interface [136].

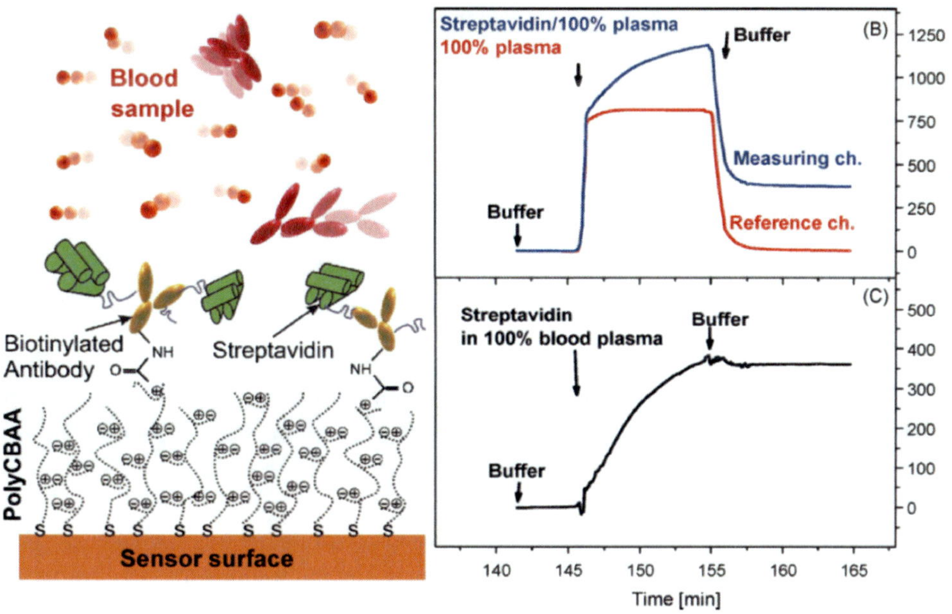

Figure 15. Sensorgrams of binding of streptavidin in un-diluted blood plasma using a polyCBAA surface, showing both measurement and reference channels and a binding schematic [129]. Reprinted from Biosensors and Bioelectronics, 24(7), Vaisocherova, H.; Zhang, Z.; Yang, W.; Cao, Z.Q.; Cheng, G.; Taylor, A.D.; Piliarik, M.; Homola, J.; Jiang, S.Y., Functionalizable surface platform with reduced non-specific protein adsorption from full blood plasma – Material selection and protein immobilization optimization, 1924-1930, Copyright 2009, with permission from Elsevier.

In seeking to detect human interferon gamma in spiked human plasma samples, different surface treatments, including various SAM coatings both with and without conjugation of a hydrogel, were examined [137]. It was found that combining dextran hydrogel with MUA gave optimal results in terms of low NSB and high specific response [137]. Other reports have advocated the use of sample pre-treatment to combat the effects of NSB from serum and plasma veterinary samples, and a precipitation technique to remove interferences is favored, involving application of saturated ammonium sulfate solution and centrifugation [138]. To detect antibodies to *Salmonella enteritidis* and *Salmonella typhimurium* in chicken serum samples, appropriate recombinant DNA antigens were immobilized and matrix effects were corrected for by use of a reference flow channel immobilized with glutathione-S-transferase [139]. Such an approach is unlikely to fully and consistently correct for NSB though.

SALIVA

Saliva offers particular challenges for use in plasmonic biosensors. Firstly, the high viscosity of the matrix makes it difficult to pass through the microfluidics of biosensors and can cause blockages. Secondly, the high concentrations of proteins, particularly mucins, mean that the surfaces can be easily fouled by adherent high mass contaminants that can greatly affect the signal from NSB to the surface in a way that often cannot be predicted. This not only diminishes sensitivity but also increases variability greatly. The situation is exacerbated by the large differences in mucin content that can occur between individuals and even separate samples of the same individual and by the propensity for mucins to precipitate out of the saliva on standing.

Perhaps the most common means of dealing with these challenges is to physically separate the high molecular mass and high viscosity components from the saliva using filters or membranes. Membrane filtration has been applied to the removal of the mucins, higher molecular weight glycoproteins, cellular material and debris before quantitation of phenytoin but it was only partially successful [140]. Use of an H-filter to extract the lower molecular weight phenytoin from the saliva proved more useful, removing 97% of the glycoproteins and reducing surface fouling by a factor of 3.6-times [140]. The disadvantage is that the analyte extraction efficiency was only 23% [140]. Another approach has been to develop an in-line filtering system for separating small molecules from large molecules in human saliva before passing over an SPR sensor for analysis of cortisol [141], Figure 16. This approach requires pre-dilution and pre-treatment of the saliva samples, has a filter that is liable to clog with time and also demonstrates limited sensitivity [141].

Saliva can be analyzed without the need for extensive pre-treatment or filtering when an appropriate additive is used in the antibody diluent for immunoassay. In the case of a plasmonic immunosensor for the steroid hormone cortisol, addition of 0.01% w/v SDS to the antibody diluent was effective in suppressing NSB to a low and consistent level, such that an immunoassay of cortisol in human saliva could be conducted with good reproducibility (inter-assay CV of 13.5%) and a low LOD (49 pg/mL), without solvent extraction or extensive pre-treatment [142]. When measuring very low concentrations of analytes in saliva, such as testosterone, the use of gold nanoparticle labeling to increase sensor signal can greatly improve results by boosting the specific signal in the biosensor relative to any NSB from the salivary matrix [143]. Concentrations of interleukin-8 (IL-8), a chemokine protein, in human saliva were detected using a BIAcore™ SPR biosensor with a sandwich immunoassay format and NSB to the sensor surface was reduced by addition of CM dextran sodium salt at very high concentrations (10 mg/mL), which is thought to compete with the surface for NSB [144].

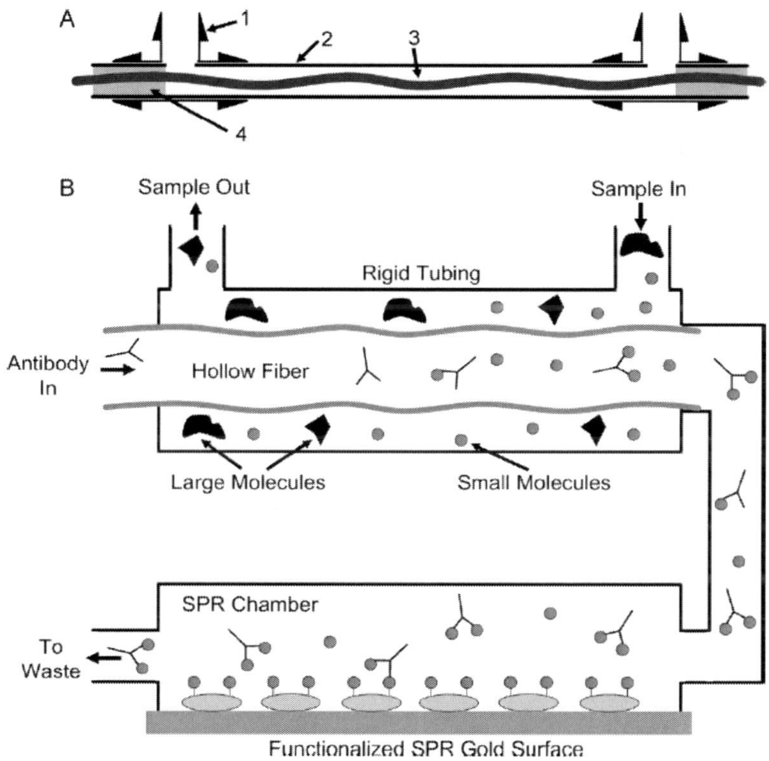

Figure 16. Use of pre-filtering in-line with an SPR flow-cell for detection of cortisol in human saliva [141]. Reprinted with permission from Stevens, R.C.; Soelberg, S.D.; Near, S.; Furlong, C.E. *Anal. Chem.* 2008, *80*, 6747-6751. Copyright 2008 American Chemical Society.

OTHER FLUIDS

A few papers have been devoted to measurement in various different media such as water samples, milk, food products and miscellaneous biological fluids, for specific applications often related to environmental monitoring or toxicology. Each medium has its own specific requirements with regards to combating NSB.

A fiber-optic SPR sensor has been applied to measurement of salinity in seawater [145]. In this case, the use of organic thiols has been proposed as a possible means of reducing fouling by adsorption of electrolytes to the sensor surface [145]. Salt deposition was postulated as the cause of high errors in the sensor response at low temperature [145]. When detecting microcystin-LR in various water samples using a total internal fluorescence format, a protein conjugate of the analyte was synthesized [146]. The sensor surface with this protein conjugate attached via a SAM of thiol-silane employing a heterobifunctional intermediate linker in a fiber optic probe then gave quite low NSB from proteins [146], likely due to the blocking behavior of the carrier protein.

Cell culture medium was the matrix of interest when SPR was applied to measurement of interleukin-6 as a potential diagnostic tool [147]. Here NSB was reduced by use of a mixed SAM composed of mercaptohexanol as blocking chain and MUA as the functionalized chain for immobilization of antibody in a sandwich immunoassay format [147]. Another interesting medium, in which an SPR biosensor has been used, is human synovial fluid where it has been applied to detection of glucose 6-phosphate isomerase (GPI) antibodies in osteoarthritis and rheumatoid arthritis patients using CM5 dextran sensor chips [148]. Recombinant human GPI proteins, with or without fusion to N utilization substance A (NusA), were immobilized on the chips, and BSA and NusA used as reference channel coatings to provide an indication of the level of NSB, thus allowing reference subtraction [148]. Immunospecificity of the interaction could be demonstrated by binding of an anti-human immunoglobulin and the NusA-fused immobilization was found to give low background NSB [148].

Milk is another challenging matrix because it's very high fat levels may interfere with availability of lipophilic analytes and cause fouling of the sensor surface. Detection of alpha-lactalbumin in milk has been achieved by direct binding to immobilized antibody and NSB has been minimized by extensive dilution of the milk samples to a final dilution of 1:10,000, which enabled detection down to 0.12 mg/mL [7]. Another study looking at detection of lactoferrin in infant formula, bovine milk and colostrums showed that the degree of sample dilution required depended upon the matrix, even to the extent of adjusting dilution for the timing of the raw milk sample collection from the cow post-partum [149]. Dilution factors of between 1:10,000 and 1:1,000 were employed and gave assay binding inhibitions of >98%, indicating low NSB on the sensor surfaces [149]. Similarly tailored dilution factors for milk samples were employed when detecting folate-binding protein in raw and commercial milks with dilution factors of between 1:50 and 1:1000 being employed and giving NSB levels of <5% of the specific response [150]. Penicillins and cephalosporins have been detected in milk, where samples demonstrated a high degree of variability in the matrix NSB across 19 individual cows, emphasizing the need for careful regulation of the NSB problem [151]. The levels of NSB were reduced to low and reasonably consistent levels by use of heat treatment and centrifugation of samples and the use of CM dextran as an additive in the samples allowing more reliable detection in individual samples as opposed to bulk defatted raw milk [151]. In an earlier study, measurement of sulphamethazine in milk samples without matrix dilution or pre-treatment was attempted but high levels of variability in the NSB responses (29 RU) were reported [152], which could have affected assay results.

In determining clenbuterol concentrations in urine and serum, a full factorial design approach has been taken to optimizing the assay and reducing NSB. To minimize matrix interferences, the authors employed ultra-filtration to physically remove interfering substances from the matrix and this was found to be effective in both urine and serum matrices. The NSB from the urine matrix was seen to be greatly affected by salt concentrations and buffer pH [153], which suggests a high electrostatic component in the interaction. The buffer composition was seen to have little effect on the level of interference to the antigen / antibody binding interaction [153]. Interestingly, antibody dilution was found to improve performance but sample dilution did not [153].

Conclusion

A great diversity of chemical architectures has been developed to improve sensitivity in SPR biosensors but such designs are of little practical use if the sensors cannot cope with the significant challenges of measuring analytes in very complex biological matrices in real-world samples. Where matrix components can either non-specifically bind to the sensor surface or interfere with antibody / antigen interactions, it is crucial to neutralize or minimize these effects whilst still retaining an ability to measure to low concentrations. Early on, the usefulness of coating gold surfaces with thiol SAMs became apparent and they have provided a convenient means of reducing NSB to bare gold surfaces. However, such coatings typically have a limited lifetime and on their own offer low immobilization capacity. Much interest has been focused on the use of polymer coatings to improve immobilization densities, with most work being done on dextran-based coatings. Whilst dextran hydrogels do definitely improve binding capacity and provide some resistance to NSB, their non-specific interactions with matrix biomolecules have proven problematic and chemical or physical techniques are needed to minimize them. Intermediate linkers have played a crucial role in improving specific binding through relief of steric constriction, whilst at the same time they can be chemically tailored to minimize NSB as in the case of OEG-based chains. An exciting new approach is the development of polymer formulations to improve upon conventional dextran brush approaches and introduce features such as adjustable immobilization density, low NSB and rapidity and ease of coating.

Each application requires its own tailored approaches to minimization of NSB, and this can be seen both for the identity of the analyte itself, e.g. DNA, and with changes in the sample medium. Some approaches have used physical separation techniques, but often the most effective approachcs have been to carefully examine the charge state and layer structure of the sensor interface and/or to employ careful adjustments to diluent conditions either through modified buffers or use of targeted additive treatments. The most successful techniques have been those that have reduced NSB to low and reproducible levels whilst retaining high binding signals and low detection limits and using chemically robust surfaces. In future, it seems likely that more attention will be focused on careful control of surface functionality combined with well-selected additives to assay diluents that can simply and affordably enable sensitive detection in biological matrices without the need for significant sample pre-treatment, pre-dilution, filtering or the use of expensive, elaborate and fragile sensor surface interfaces.

REFERENCES

[1] Cooper, M. A. *Nat. Rev. Drug Discov.* 2002, *1*, 515-528.

[2] Myszka, D.G.; Rich, R.L. *Pharm. Sci. Technol. To.* 2000, *3*, 310-317.

[3] Marchesini, G.R.; Koopal, K.; Meulenberg, E.; Haasnoot, W.; Irth, H. *Biosens. Bioelectron.* 2007, *22*, 1908-1915.

[4] Lotierzo, M.; Henry, O.Y.F.; Piletsky, S.; Tothill, I.; Cullen, D.; Kania, M.; Hock, B.; Turner, A.P.F. *Biosens. Bioelectron.* 2004, *20*, 145-152.

[5] Homola, J. *Chem. Rev.* 2008, *108*, 462-493.

[6] Hendry, E.; Carpy, T.; Johnston, J.; Popland, M.; Mikhaylovskiy, R.V.; Lapthorn, A.J.; Kelly, S.M.; Barron, L.D.; Gadegaard, N.; Kadodwala, M. *Nat. Nanotechnol.* 2010, *5*, 783-787.

[7] Indyk, H.E. *Int. Dairy J.* 2009, *19*, 36-42.

[8] Mizuta, Y.; Onodera, T.; Singh, P.; Matsumoto, K.; Miura, N.; Toko, K. *Biosens. Bioelectron.* 2008, *24*, 191-197.
[9] Taylor, A.D.; Ladd, J.; Yu, Q.M.; Chen, S.F.; Homola, J.; Jiang, S.Y. *Biosens. Bioelec.* 2006, *22*, 752-758.
[10] Ishii, T.; Suzuki, Y.; Akiyama, Y.; Otsuka, H.; Kataoka, K.; Nagasaki, Y. *Kobunshi Robunshu* 2005, *62*, 81-86.
[11] Choi, S; Chae, J. *J. Micromech. Microeng.* 2010, *20*, 075015.
[12] Anderson, G.P.; Merrick, E.C.; Trammell, S.A.; Chinowsky, T.M.; Shenoy, D.K. *Sens. Lett.* 2005, *3*, 151-156.
[13] Cao, C.; Kim, J.P.; Kim, B.W.; Chae, H.; Yoon, H.C.; Yang, S.S.; Sim, S.J. *Biosens. Bioelctron.* 2006, *21*, 2106-2113.
[14] Stapleton, S.; Bradshaw, B.; O'Kennedy, R. *Anal. Chim. Acta* 2009, *651*, 98-104.
[15] Mar, M.N.; Ratner, B.D.; Yee, S.S. *Sensor Actuat. B-Chem.* 1999, *54*, 125-131.
[16] Dalsin, J.L.; Messersmith, P.B. *Mater. Res. Soc. Symp. P* 2003, *774*, 75-80.
[17] Krishnamoorthy, G.; Carlen, E.T.; van der Berg, A.; Schasfoort, R.B.M. *Sensor Actuat. B-Chem.* 2010, *148*, 511-521.
[18] Fontana, E.; Dutra, R.F.; Neto, J.O.M. *in* International Microwave and Optoelectronics Conference, 2001, 181-184.
[19] Homola, J.; Lu, H.B.B.; Nenninger, G.G.; Dostelek, J.; Yee, S.S. *Sensors Actuat. B-Chem.* 2001, *76*, 403-410.
[20] Chung, J.W.; Bernhardt, R.; Pyun, J.C. *J. Immunol. Methods* 2006, *311*, 178-188.
[21] Birnbaumer, G.M.; Lieberzeit, P.A.; Richter, L.; Schirhagl, R.; Milnera, M.; Dickert, F.L.; Bailey, A.; Ertl, P. *Lab Chip* 2009, *9*, 3549-3556.
[22] Grasso, G.; D'Agata, R.; Zanoli, L.; Spoto, G. *Microchem. J.* 2009, *93*, 82-86.
[23] Wong, C.L.; Ho, H.P.; Lei, K.F.; Li, W.J.; Chan, K.S.; Law, W.C.; Wu, S.Y.; Kong, S.K.; Lin, C.L. *IEEE C Elec. Devices* 2005, 709-712.
[24] Karp, N.A.; Edwards, P.R.; Leatherbarrow, R.J. *Biosens. Bioelectron.* 2005, *21*, 128-134.
[25] Brink, G.; Sigl, H.; Sackmann, E. *Sensor Actuat. B-Chem.* 1995, *25*, 756-761.
[26] Magalhaes, A.; Bordeira, S.; Almeida, A.C.; Fontes, V.; Costa, M.J.L.; Fonseca, L.P.; da Fonseca, J.G. *P. Soc. Photo.-Opt. Ins.* 2009, *7169*, 716903.
[27] Naimushin, A.N.; Soelberg, S.D.; Nguyen, D.K.; Dunlap, L.; Bartholomew, D.; Elkind, J.; Melendez, J.; Furlong, C.E. *Biosens. Bioelectron.* 2002, *17*, 573-584.
[28] Nedelkov, D.; Nelson, R.W. *Biosens. Bioelectron.* 2001, *16*, 1071-1078.
[29] Nedelkov, D.; Rasooly, A.; Nelson, R.W. *Int. J. Food Microbiol.* 2000, *60*, 1-13.
[30] Piehler, J.; Brecht, A.; Geckeler, K.E.; Gauglitz, G. *Biosens. Bioelctron.* 1996, *11*, 579-590.
[31] Vikholm-Lundin, I.; Pulli, T.; Albers, W.M.; Tappura, K. *Biosens. Bioelectron.* 2008, *24*, 1036-1038.
[32] Kim, E.; Lee, B.S.; Pyo, H.B.; Song, H.W.; Kim, Y.P.; Choi, I.S.; Kim, H.S. *Biochip J.* 2008, *2*, 103-110.
[33] Carrigan, S.D.; Scott, G.; Tabrizian, M. *Biomaterials* 2005, *26*, 7514-7523.
[34] Vestergaard, M.; Kernman, K.; Kim, D.K.; Hiep, H.M.; Tamiya, E. *Talanta* 2008, *74*, 1038-1042.
[35] Vikholm-Lundin, I.; Albers, W.M. *Biosens. Bioelectron.* 2006, *21*, 1141-1148.

[36] Karamanska, R.; Mukhopadhyay, B.; Russell, D.A.; Field, R.A. *Chem. Commun.* 2005, *26*, 3334-3336.
[37] Redman, J.E. *Methods* 2007, *43*, 302-312.
[38] Masson, J.F.; Battaglia, T.M.; Beaudoin, S; Booksh, S. *P. Soc. Photo.-Opt. Ins.* 2004, *5588*, 178-191.
[39] Avilov, S.V.; Aleksandrova, N.A.; Kunda, Y.M.; Verevka, S.V.; Shyrshov, I.M. *Ukr. Biokhim. Zh+.* 2004, *76*, 98-103.
[40] Perelman, L.A.; Schwartz, M.P.; Wohlrab, A.M.; VanNieuwenhze, M.S.; Sailor, M.J. *Phys. Status Solidi A* 2007, *204*, 1394-1398.
[41] Schwartz, M.P.; Yu, C.; Alvarez, S.D.; Migliori, B.; Godin, D.; Chao, L.; Sailor, M.J. *Phys. Status Solidi A* 2007, *204*, 1444-1448.
[42] Hu, W.P.; Hsu, H.Y.; Chiou, A.; Tseng, K.Y.; Lin, H.Y.; Chang, G.L.; Chen, S.J. *Biosens. Bioelectron.* 2006, *21*, 1631-1637.
[43] Chin, S.F.; Pantano, P. *Microchem. J.* 2006, *84*, 1-9.
[44] Liu, Z.; Amridis, M.D. *Surf. Sci.* 2005, *596*, 117-125.
[45] Shelver, W.L.; Keum, Y.S.; Li, Q.X.; Fodey, T.L.; Elliott, C.T. *Food Agric. Immunol.* 2005, *16*, 199-211.
[46] Neff, H.; Beeby, T.; Lima, A.M.N.; Borre, M.; Thirstrup, C.; Zong, W.; de Almeida, L.A.L. *Biosens. Bioelectron.* 2006, *21*, 1746-1752.
[47] Li, L.Y.; Chen, S.F.; Jiang, S.Y. *J. Biomat. Sci.-Polym. E* 2007, *18*, 1415-1427.
[48] Otsuka, H.; Nagasaki, Y.; Kataoka, K. *Adv. Drug Deliv. Rev.* 2003, *55*, 403-419.
[49] Nogues, C.; Leh, H.; Langendorf, C.G.; Law, R.H.P.; Buckle, A.M.; Buckle, M. *PLoS One* 2010, *5*, e12152.
[50] Kitano, H.; Kondo, Y.; Saito, D.; Gemmei-Ide, M.; Morita, H.; Kanayama, N. *Colloid Surface B* 2010, *78*, 155-162.
[51] Jeong, J.; Lee, C.S.; Chung, S.J.; Chung, B.H. *Bioproc. Biosyst. Eng.* 2010, *33*, 165-169.
[52] Kyprianou, D.; Guerreiro, A.R.; Nirschl, M.; Chianella, I.; Subrahmanyam, S.; Turner, A.P.F.; Piletsky, S. *Biosens. Bioelctron.* 2010, *25*, 1049-1055.
[53] Park, T.J.; Hyun, M.S.; Lee, H.J.; Lee, S.Y.; Ko, S. *Talanta* 2009, *79*, 295-301.
[54] Cho, H.M.; Jang, D.H.; Lee, S.K.; Ku, S.J.; Kim, H.C.; Yu, K.S.; Klee, J.H. *Biochip. J.* 2008, *2*, 206-211.
[55] Kim, H.; Noh, J.; Hara, M.; Lee, H. *Ultramicroscopy* 2008, *108*, 1140-1143.
[56] Klein, E.; Kerth, P.; Lebeau, L. *Biomaterials* 2008, *29*, 204-214.
[57] Kawaguchi, T.; Shankaran, D.R.; Kim, S.J.; Gobi, K.V.; Matusmoto, K.; Toko, K.; Miura, N. *Talanta* 2007, *72*, 554-560.
[58] Nagatomo, K.; Kawaguchi, T.; Miura, N.; Toko, K.; Matsumoto, K. *Talanta* 2009, *79*, 1142-1148.
[59] Tappura, K.; Vikholm-Lundin, I.; Albers, W.M. *Biosens. Bioelectron.* 2007, *22*, 912-919.
[60] Huang, L.; Reekmans, G.; Saerens, D.; Friedt, J.M.; Frederix, F.; Francis, L.; Muyldermans, S.; Campitelli, A.; van Hoof, C. *Biosens. Bioelectron.* 2005, *21*, 483-490.
[61] Choi, S.H.; Lee, J.W.; Sim, S.J. *Biosens. Bioelectron.* 2005, *21*, 378-383.
[62] Jung, L.S.; Nelson, K.E.; Campbell, C.T.; Stayton, P.S.; Yee, S.S.; Perez-Luna, V.; Lopez, G.P. *Sensor Actuat. B-Chem.* 1999, *54*, 137-144.

[63] Kim, H.; Cho, I.H.; Park, J.H.; Kim, S.; Paek, S.H.; Noh, J.; Lee, H. *Colloid Surface A* 2008, *313*, 541-544.
[64] Lahiri, J.; Isaacs, L.; Tien, J.; Whitesides, G.M. *Anal. Chem.* 1999, *71*, 777-790.
[65] Gobi, K.V.; Iwasaka, H.; Miura, N. *Biosens. Bioelectron.* 2007, *22*, 1382-1389.
[66] Yam, C.M.; Zheng, L.; Salmain, M.; Pradier, C.M.; Marcus, P.; Jaouen, G. *Colloid Surface B* 2001, *21*, 317-327.
[67] Gao, C.L.; Li, G.Z.; Xue, H.; Yang, W.; Zhang, F.B.; Jiang, S.Y. *Biomaterials*, 2010, *31*, 1486-1492.
[68] Chen, H.; Zheng, Y.; Jiang, J.H.; Wu, H.L.; Shen, G.L.; Yu, R.Q. *Biosens. Bioelectron.* 2008, *24*, 684-689.
[69] Luckarift, H.R.; Balasubramanian, S.; Paliwal, S.; Johnson, G.R.; Simonian, A.L. *Colloid Surface B* 2007, *58*, 28-33.
[70] Gobi, K.V.; Kim, S.J.; Tanaka, H.; Shoyama, Y.; Miura, N. *Sensor Actuat. B-Chem.* 2007, *123*, 583-593.
[71] Dutra, R.F.; Kubota, L.T. *Clin. Chim. Acta* 2007, *376,* 114-120.
[72] Lee, C.S.; Lee, S.H.; Kim, Y.G.; Lee, J.H.; Kim, Y.K.; Kim, B.G. *Biosens. Bioelectron.* 2007, *22*, 891-898.
[73] Bucur, B.; Danet, A.F.; Marty, J.L. *Biosens. Bioelectron.* 2004*, 20*, 217-225.
[74] Liu, Z.; Amiridis, M.D. *Colloid Surface B* 2004, *35*, 197-203.
[75] Baek, S.H.; Shin, Y.B.; Kim, M.G.; Ro, H.S.; Kim, E.K.; Chung, B.H. *Biotechnol. Bioproc. E* 2004, *9*, 143-146.
[76] Disley, D.M.; Cullen, D.C.; You, H.X.; Lowe, C.R. *Biosens. Bioelectron.* 1998, *13*, 1213-1225.
[77] Morgan, H.; Taylor, D.M.; D'Silva, C. *Thin Solid Films* 1992, *209* 122-126.
[78] Vikholm, I.; Viitala, T.; Albers, W.M.; Peltonen, J. *BBA-Biomembranes* 1999, *1421*, 39-52.
[79] Webster, C.I.; Cooper, M.A.; Packman, L.C.; Williams, D.H.; Gray, J.C. *Nucleic Acids Res.* 2000, *28*, 1618-1624.
[80] Piehler, J.; Brecht, A.; Hehl, K.; Gauglitz, G. *Colloid Surface B* 1999, *13*, 325-336.
[81] Masson, J.F.; Battaglia, T.M.; Davidson, M.J.; Kim, Y.C.; Prakash, A.M.C.; Beaudoin, S.; Booksh, K. *Talanta* 2005, *67*, 918-925.
[82] Masson, J.F.; Battaglia, T.M.; Kim, Y.C.; Prakash, A.; Beaudoin, S.; Booksh, K. *Talanta* 2004, *64*, 716-725.
[83] Fong, C.C.; Wong, M.S.; Fong, W.F.; Yang, M.S. *Anal. Chim. Acta* 2002 *456*, 201-208.
[84] Brault, N.D.; Gao, C.L.; Xue, H.; Piliark, M.; Homola, J.; Jiang, S.Y.; Yu, Q.M. *Biosens. Bioelectron.* 2010, *25*, 2276-2282.
[85] Yoshimoto, K.; Hirase, T.; Madsen, J.; Armes, S.P.; Nagasaki, Y. *Macromol. Rapid Comm.* 2009, *30*, 2136-2140.
[86] Bergstrand, A.; Rahmani-Monfared, G.; Ostland, A.; Nyden, M.; Holmberg, K. *J. Biomed. Mater. Res. A*, 2009, *88A*, 608-615.
[87] Thompson, L.B.; Mack, N.H.; Nuzzo, R.G. *Phys. Chem. Chem. Phys.* 2010, *12*, 4301-4308.
[88] Jeong, H.J.; Pyun, W.H.; Yang, S.Y. *Macromol. Rapid Comm.* 2009, *30*, 1109-1115.
[89] Zhen, G.L.; Falconnet, D.; Kuennemann, E.; Voros, J.; Spencer, N.D.; Textor, M.; Zurcher, S. *Adv. Funct. Mater.* 2006, *16*, 243-251.

[90] Disley, D.M.; Blyth, J.; Cullen, D.C.; You, H.X.; Eapen, S.; Lowe, C.R. *Biosens. Bioelectron.* 1998, *13*, 383-396.
[91] Mi, L.; Bernards, M.T.; Cheng, G.; Yu, Q.M.; Jiang, S.Y. *Biomaterials* 2010, *31*, 2919-2925.
[92] Chang, Y.; Chu, W.L.; Chen, W.Y.; Zheng, J.; Liu, L.Y.; Ruaan, R.C.; Higuchi, A. *J. Biomed. Mater. Res. A* 2010, *93A*, 400-408.
[93] Kurita, R.; Hirata, Y.; Yabuki, S.; Yokota, Y.; Kato, D.; Sato, Y.; Mizutani, F.; Niwa, O. *Sensor Actuat. B-Chem.* 2008, *130*, 320-325.
[94] Kurita, R.; Hirata, Y.; Yabuki, S.; Kato, D.; Sato, Y.; Mizutani, F.; Niwa, O. *Electrochem.* 2006, *74*, 121-124.
[95] Piehler, J.; Brecht, A.; Vailokas, R.; Liedberg, B.; Gauglitz, G. *Biosens. Bioelectron.* 2000, *15*, 473-481.
[96] Chang, Y.; Shu, S.H.; Shih, Y.J.; Chu, C.W.; Ruaan, R.C.; Chen, W.Y. *Langmuir* 2010, *26*, 3522-3530.
[97] Mrabet, B.; Nguyen, M.N.; Majbri, A.; Mahouche, S.; Turmine, M.; Bakhrouf, A.; Chehimi, M. M. *Surf. Sci.* 2009, *603*, 2422-2429.
[98] Zhang, Z.; Chen, S.F.; Chang, Y.; Jiang, S.Y. *J. Phys. Chem. B* 2006, *110*, 10799-10804.
[99] Mizukami, K.; Takakura, H.; Matusnaga, T.; Kitano, H. *Colloid Surface B* 2008, *66*, 110-118.
[100] Cheng, X. H.; Canavan, H.E.; Graham, D.J.; Castner, D.G.; Ratner, B.D. *Biointerfaces* 2006, *1*, 61-72.
[101] Chu, L.Q.; Knoll, W.; Forch, R. *Biosens. Bioelectron.* 2009, *25*, 519-522.
[102] Hook, A.L.; Thissen, H.; Voelcker, N.H. *Langmuir* 2009, 9173-9181.
[103] Aulasevich, A.; Roskamp, R.F.; Jonas, U; Menges, B.; Dostalek, J.; Knoll, W. *Macromol. Rapid Comm.* 2009, *30*, 872-877.
[104] Shang, F.J.; Liu, Y.L.; Hrapovic, S.; Glennon, J.D.; Luong, J.H.T. *Analyst* 2009, *134*, 519-527.
[105] Li, G.Z.; Cheng, G.; Xue, H.; Chen, S.F.; Zhang, F.B.; Jiang, S.Y. *Biomaterials* 2008, *29*, 4592-4597.
[106] Chi, Y.S.; Lee, B.S.; Kil, M.; Jung, H.J.; Oh, E.; Choi, I.S. *Chem-Asian J.* 2009, *4*, 135-142.
[107] Kyprianou, D.; Guerreiro, A.R.; Chianella, I.; Piletska, E.V.; Fowler, S.A.; Karim, K.; Whitcombe, M.J.; Turner, A.P.F.; Piletsky, S.A. *Biosens. Bioelectron.* 2009, *24*, 1365-1371.
[108] Iwasaki, Y.; Takami, U.; Shinohara, Y.; Kurita, K.; Akiyoshi, K. *Biomacromolecules* 2007, *8*, 2788-2794.
[109] Park, J.; Kurosawa, S.; Takai, M.; Ishihara, K. *Colloid Surface B* 2007, *55*, 164-172.
[110] Vikholm-Lundin, I.; Piskonen, R.; Albers, W.M. *Biosens. Bioelectron.* 2007, *22*, 1323-1329.
[111] Vikholm, I. *Sensor Actuat. B-Chem.* 2005, *106*, 311-316.
[112] Kitano, H.; Gemmei-Ide, M.; Anraku, Y.; Saruwatari, Y. *Colloid Surface B* 2007, *56*, 188-196.
[113] Zhou, Y.; Andersson, O.; Lindberg, P.; Liedberg, B. *Microchim. Acta* 2004, *147*, 21-30.
[114] Forde, G.M.; Ghose, S.; Slater, N.K.H.; Hine, A.V.; Darby, R.A.J.; Hitchcock, A.G. *Biotechnol. Bioeng.* 2006, *95*, 67-75.

[115] Mir, M.; Cameron, P.J.; Zhong, X.H.; Azzaroni, O.; Alvarez, M.; Knoll, W. *Talanta* 2009, *78*, 1102-1106.
[116] Vikholm-Lundin, I.; Piskonen. R. *Sensor Actuat. B-Chem.* 2008, *134*, 189-192.
[117] Hajdukiewicz, J.; Boland, S.; Kavanagh, P.; Leech, D. *Biosens. Bioelectron.* 2010, *25*, 1037-1042.
[118] Pollet, J.; Delport, F.; Janssen, K.P.F.; Jans, K.; Maes, G.; Pfeiffer, H.; Wevers, W.; Lammertyn, J. *Biosens. Bioelectron.* 2009, *25*, 864-869.
[119] Park, K.H.; Kim, S.; Yang, S.M.; Park. H.G. *J. Nanosci. Nanotechno.* 2009, *9*, 1374-1378.
[120] Boireau, W.; Zeeh, J.C.; Puig, P.E.; Pompon, D. *Biosens. Bioelectron.* 2005, *20*, 1631-1637.
[121] Shumaker-Parry, J.S.; Campbell, C.T.; Stormo, G.D.; Silbaq, F.S.; Aebersold, R.H. *P. Soc. Photo.-Opt. Ins.* 2000, *1*, 158-166.
[122] Wang, J.L.; Munir, A.; Li, Z.H.; Zhou, H.S. *Biosens. Bioelctron.* 2009, *25*, 124-129.
[123] Win, M.N.; Klein, J.S.; Smolke, C.D. *Nucleic Acids Res.* 2006, *34*, 1653-1665.
[124] Otamiri, M.; Nilsson, K.G.I. *Int. J. Biol. Macromol.* 1999, *26*, 263-268.
[125] Quinn, J.G.; O'Kennedy, R.; Smyth, M.; Moulds, J.; Frame, T. *J. Immunol. Methods* 1997, *206*, 87-96.
[126] Emmenegger, C.R.; Brynda, E.; Riedel, T.; Sedlakova, Z.; Houska, M.; Alles, A.B. *Langmuir* 2009, *25*, 6328-6333.
[127] Piliarik, M.; Bockova M.; Homola, J. *Biosens. Bioelectron.* 2010, *26*, 1656-1661.
[128] Bolduc, O.R.; Pelletier, J.N.; Masson, J.F. *Anal. Chem.* 2010, *82*, 3699-3706.
[129] Vaisocherova, H.; Zhang, Z.; Yang, W.; Cao, Z.Q.; Cheng, G.; Taylor, A.D.; Piliarik, M.; Homola, J.; Jiang, S.Y. *Biosens. Bioelectron.* 2009, *24*, 1924-1930.
[130] Vaisocherova, H.; Yang, W.; Zhang, Z.; Cao, Z.Q.; Cheng, G.; Piliarik, M.; Homola, J.; Jiang, S.Y. *Anal. Chem.* 2008, *80*, 7894-7901.
[131] Monchaux, E.; Vermette, P. *Langmuir* 2007, *23*, 3290-3297.
[132] Situ, C.; Wylie, A.R.G.; Douglas, A.; Elliott, C.T. *Talanta* 2008, *76*, 832-836.
[133] Phillips, K.S.; Han, J.H.; Cheng, Q. *Anal. Chem.* 2007, *79*, 899-907.
[134] Trevino, J.; Calle, A.; Rodriguez-Frade, J.M.; Mellado, M.; Lechuga, L.M. *Talanta* 2009, *78*, 1011-1016.
[135] Keating, G.J.; Quinn, J.G.; O'Kennedy, R. *Anal. Lett.* 1999, *32*, 2163-2176.
[136] Ayela, C.; Roquet, F.; Valera, L.; Granier, C.; Nicu, L.; Pugniere, M. *Biosens. Bioelctron.* 2007, *22*, 3113-3119.
[137] Stigter, E.C.A.; de Jong, G.J.; van Bennekom, W.P. *Biosens. Bioelectron.* 2005, *21*, 474-482.
[138] Johnsson, L.; Baxter, G.A.; Crooks, S.R.H.; Brandon, D.L.; Elliott, C.T. *Food Agric. Immunol.* 2002, *14*, 209-216.
[139] Jongerius-Gortenmaker, B.G.M.; Goverde, R.L.J.; van Knapen, F.; Bergwerff, A.A. *J. Immunol. Methods* 2002, *266*, 33-44.
[140] Helton, K.L.; Nelson, K.E.; Fu, E.; Yager, P. *Lab Chip* 2008, *8*, 1847-1851.
[141] Stevens, R.C.; Soelberg, S.D.; Near, S.; Furlong, C.E. *Anal Chem* 2008, *80*, 6747-6751.
[142] Mitchell, J.S.; Lowe, T.E.; Ingram, J.R. *Analyst* 2009, *134*, 380-386.
[143] Mitchell, J.S.; Lowe, T.E. *Biosens. Bioelctron.* 2009, *24* 2177-2183.
[144] Yang, C.Y.; Brooks, E.; Li, Y.; Denny, P.; Ho, C.M.; Qi, F.X.; Shi, W.Y.; Wolinsky, L.; Wu, B.; Wong, D.T.W.; Montemagno, C.D. *Lab Chip* 2005, *5*, 1017-1023.

[145] Gentleman, D.J.; Booksh, K.S. *Talanta* 2006, *68*, 504-515.
[146] Long, F.; He., M.; Zhu, A.N.; Shi, H.C. *Biosens. Bioelectron.* 2009, *24*, 2346-2351.
[147] Chou, T.H.; Chuang, C.Y.; Wu, C.M. *Cytokine*, 2010, *51*, 107-111.
[148] Kim, J.Y.; Lee, M.H.; Jung, K.I.; Na, H.Y.; Cha, H.S.; Ko, E.M.; Kim, T.J. *Exp. Mol. Med.* 2003, *35*, 310-316.
[149] Indyk, H.E.; Filonzi, E.L. *Int. Dairy J.* 2005, *15*, 429-438.
[150] Indyk, H.E.; Filonzi, E.L. *J. Agr. Food Chem.* 2004, *52*, 3253-3258.
[151] Cacciatore, G.; Petz, M.; Rachid, S.; Hakenbeck, R.; Bergwerff, A.A. *Anal. Chim. Acta* 2004, *520*, 105-115.
[152] Bergstrom, C.; Sternesjo, A.; Bjurling, P.; Lofas, S. *Food Agric. Immunol.* 1999, *11*, 329-338.
[153] Johansson, M.A.; Hellenas, K.E. *Analyst* 2004, *129*, 855-859.

In: Plasmons: Structure, Properties and Applications
Editors: Anton E. Turunen and Johannes O. Niemi
ISBN: 978-1-62100-060-0

Chapter 2

DESIGN AND CONSTRUCTION OF MULTIFUNCTIONAL SERS-ACTIVE SUBSTRATES FOR BIOELECTRONIC DEVICES

Jiu-Ju Feng[1,2], Ai-Jun Wang[1], Sezer Murat[2], Inez M. Weidinger[2] and Peter Hildebrandt[2]

[1] School of Chemistry and Environmental Science,
Henan Normal University, Xinxiang, Henan, China
[2] Technische Universität Berlin, Institut für Chemie, Germany

ABSTRACT

This chapter discusses the design and construction of surface-enhanced Raman scattering (SERS) active nanostructures with tunable size-/shape- dependent physical and chemical properties, and provides examples of exploit their wide applications in many scientific and technological fields. Coupled with biomolecules, different hybrid devices were developed, which offered a powerful platform to study the conformation change, dynamic behavior and biocatalysis of the target redox proteins like cytochrome c (Cyt-c, usually as a model protein). It would be benefit to deeply understand the relationship of the structure and function in biological system. Moreover, these hybrid devices could serve as an optical handle to investigate the properties of the structures themselves, such as how to control their optic-/electric- response for spectroelectrochemistry, which are widely used in (bio)analytical chemistry and interfacial science. In this field, the (electro)depositon of novel SRES-active (bio)structures directly on electrodes become a highlight and exciting goal, which produces inexpensive, controllable, reproducible, reusable biointerfaces by tailoring the morphology of the deposits, biomolecules, or combined both together.

This chapter reviewed recent advances in the construction and applications of the biomaterials, mainly contributed by our laboratory in the latest five years. We roughly describe several typical optobioelectronic devices, which combine biotechnology with optical (eg. surface enhanced resonance Raman scattering, SERRS) and electrochemical techniques, while the rapid progress of the nanotechnology provides extra opportunities for investigations in this field. Firstly, we discussed the recent development of the SERS-active nanostructures. These nanostructures are functionalized with biocompatible coatings focused on different mecaptoalkanes for efficient immobilization of the target protein. Then, these biosensing devices were applied to study the conformation change, redox behavior and biocatalysis of the immobilized proteins. Besides, the electron transfer mechanism of the protein with the conventional electrode was also discussed, according to the experiments and interfacial bioelectrochemistry theory. Furthermore, the comparable results, obtained from spectro-electrochemistry experiments, are explored and discussed in some detail. Finally, some significant advances combined electrochemistry with surface-sensitive SERS for the nanostructured biointerfaces are also outlined.

1. Introduction

Remarkable progresses in metal structure research have been made in the past several decades [1,2], which involved in a variety of scientists including physicists, chemists, materials scientists and biologists. Many of these applications are inspired by unique optical, electric, or catalytic properties, which are determined by size, shape, composition, and surrounding environment. Particularly important are structured noble metals, such as Au and Ag that are widely used in optics [1] and spectroscopy [3-5], nanotechnology [6-8], nanoscale photonics [9,10], scanning microscopy [11], and, most importantly, bioelectronics devices [12,13].

The ideal bioelectronics devices require immobilization of the target proteins with native structures, efficient electronic communication, and a high protein load. The surface enhanced Raman scattering (SERS) offers a strong surface plasmon resonance enhancement that allows for in-situ structural characterization of the loaded protein, which is a prerequisite for the rational improvement of the bioelectronics devices. The promising and wide applications of the noble substrates are attributed to the localized surface plasmon resonance absorption bands, a collective oscillation of the electron gas that occurs upon resonant coupling with incident electromagnetic radiation. The SERS enhancement intensity can be adjusted by varying their size, morphology, composition, and local dielectric environment in the close vicinity of the metallic surfaces.

Up to now, tremendous advances have been made to selectively enhance specific parts of the adsorbed molecules, such as protein cofactors, and thereby allow a detailed analysis of surface-bound species on the molecular level. Moreover, combined with the molecular resonance Raman (RR) effect, a new approach, that is, surface enhanced resonance Raman spectroscopy (SERRS) has been developed. This technique greatly benefits from the match of the molecular and surface plasmon resonances. Besides, the plasmon resonance also serves as an optical approach to study the properties of the structures themselves, which are widely used in (bio)analytical chemistry and interfacial science. In addition, the SERS-active conducting substrate can be used as a working electrode in electrochemistry study through using an external circuit.

Heme proteins are important redox proteins that contain the porphyrin complex of iron(II) or hemin(III) as a prosthetic group. Upon the protein binding to the conducting surfaces, the interactions between them can heavily affect the structure and function of the immobilized protein. Therefore, immobilization should be conducted using relatively mild interactions that are sufficient to keep the protein on the surface, while not change the protein structures. Secondly, shortening the distance from the active center of the protein to the substrates is essential for efficient electron communication. In the past, redox reactions of these proteins were usually probed by mediators, which transfer electrons from protein to the conventional electrode. Among them, cytochrome c (Cyt-c), a small soluble heme protein, is an ideal biomolecules for fundamental study about the redox behavior of the proteins in biological systems, as well as for biosensing and biocatalysis [3,4,14]. Cyt-c shows a positively charged lysine rich domain near the heme active center, which binds to negative-charged surfaces through electrostatic interactions. However, direct electron transfer (DET) from the protein to the conventional electrode is usually difficult to achieve, owing to the structural changes of the protein. Thus, many efforts have been made for efficient adsorption of the target protein [4,5,13,15,16]. More interestingly, many novel and advanced biomaterials have been prepared and widely applied to optical/electric electronics devices in the last decade.

Among them, the key part is design, construction and characterization of the biointerfaces. For spectro-electrochemistry devices, the SERS-active materials have to be deposited on conducting substrates. In the simplest way, the substrate itself offers Raman-active sites that can be realized by direct electrochemical roughening of the bulk material [17-19]. The surface enhancement was firstly observed on the roughened Ag electrode as continuous active metal substrate [17], for its superior electric field enhancement at the metal surfaces via the formation of surface Plasmon polaritons. However, the bare Ag surface is instable and toxic, which is prone to cause protein denaturation and decreasing the signal intensity with time [20,21]. Thus, versatile strategies were developed for the construction of biointerfaces with good biocompatible, long-term stability, fast electron communication, high SERS enhancement and catalytic efficiency [5,13,14,22]. The firstly extensively investigated system is the Ag rough electrode with different ω-functionalized mercaptoalkanes by self-assembly or covalent linkage previously [3,5,14,23]. Alternatively, other approaches were developed to construct a variety of Raman active substrates, such as lithographic methods [24-26] and (electro)deposition [27-31]. However, in the former case, expensive and large instruments are required, although this method has good repeatability. Nevertheless, the later always receives popular owing to its facile operation, precise control, low-cost, green, and good repeatability.

As known, further applications of Ag based substrates concerning spectroelectrochemistry investigations and life science are confined, because of its lower redox potential and unavoidable formation of Ag^+ ions upon exposing in aqueous solution. The generated Ag^+ ions can destroy or even degradation of the protein membrane. However, the Ag displays a broad optical enhancement profile from 400 nm to near infrared region. Although Au-derived materials are usually used in electrochemical biosensors for its wide potential window, good stability and biocompatibility, while, its optical range is restricted to the excitation lines above 600 nm. Therefore, it is impossible to study heme proteins that have distinguished absorption peaks in blue and violet region, where d–d absorptions effectively damp the formation of surface plasmon resonances. Take all together, great efforts have been focused on design and synthesis of novel optical/electric devices based on a variety of nanostructured hybrids by Ag in combination with Au [29,32-35].

To date, many SERS-active devices have been developed and become a hot project in material science and Raman field [2,15,32,36-40], which exhibit potential applications by combining the optical properties of Ag with the (electro)chemical advantages provided by Au surfaces. There are quite a lot of reports about this target, such as core/shell Au-Ag structures

[41-43], and Au (or Ag) directly deposited on the conducting substrates [35,44]. In general, it is easy to construct the core/shell hybrid structures with different noble metals, whereas it is not good to preserve their respective original properties. Particularly, great effects would impose on the original properties of the core part. For instance, the Raman signal would be greatly decreased if the Raman active metals such as Au directly deposited on the top of Raman non-active metallic surface. Conversely, a thin dielectric layer between different metals modulates light intensity, due to the interference effect [2,37,45-47]. The resulting devices have the ability to tune both the extinction spectrum and the SERS enhancement factor of the nanostructures via adjusting the thickness of the dielectric layer. The hybrid structures have synergic effect to increase the SERS enhancement factor of the nanostructures, compared to those without the dielectric layer.

In this chapter, we concentrate on design, construction, and applications of the bioelectronics devices based on noble metal structures, which mainly combine nanotechnology with optical (i.e. surface enhanced resonance Raman scattering, SERRS) and electrochemical technique, as well as biotechnology. Firstly, we discuss the recent advances of the biomaterials derived from redox protein combined with noble metal structures. Then, these biomaterials are employed as biosensing devices including biosensors or bio-fuel cells. The conformation change, redox behavior, electron transfer mechanism, and biocatalyst of the immobilized proteins are particularly addressed in some detail. Next, the remote-control optical/electric response of the typical protein of Cyt-c are explored and discussed as a model probe. Finally, different bioelectronics devices are presented and their respective advantages and disadvantages are compared, respectively.

This goal is to describe how the optical/electric signal enhancement of the target protein is affected by the size, shape and composition of the structures, and the properties of the environment. A significant emphasis will be placed on electron transfer mechanism of the protein with the electrode on different hybrid devices. This is a relatively interesting area of investigation that can offer unique information about electron dynamics of the immobilized protein and how the environment affects energy relaxation.

2. Construction of SERS-Active Substrates

Surface enhancement Raman spectroscopy, often abbreviated SERS, is a surface-sensitive technique that offers strong enhanced Raman signals from Raman-active analyte molecules that have been adsorbed onto specially prepared metallic surfaces. The first example was performed by Fleischman and coworkers using pyridine as a probe adsorbed on the roughened Ag electrode in 1974. In 1977, Van Duyne and Creighton groups independently proposed a mechanism for the observed enhancement, which are still accepted today as explaining the SERS effect. The design and construction of metallic nanostructures based on SERS effect attracted great interests in material science and Raman field [5,9,22,25,32,35,48-57], mainly because of their unique optical, electronic, and catalytic properties.

2.1. Noble Metal Nanostructures

Recently, many metal structures are fabricated, which serve as SERS-active surfaces and have wide applications [47], including in-situ and ambient analysis of electrochemical, catalytic, biological, and organic interfacial systems, owing to its surface selectivity and high sensitivity.

Au nanoflake films were constructed by using a scaffold of flake-built spherical dipeptide aggregations as templates [58]. Many shape-tailored and surfactant-assisted gold nanocrystals were constructed with the assistance of surfactants by the emerging colloidal techniques [59]. Ag nanostructures were in situ generated through citrate reduction in the ultra-structure of the sporopollenin biopolymer [60]. Large-area, freestanding, two-dimensionally ordered macroporous Ag thin films with adjustable periodic spacings and good mechanical robustness were constructed using nanosphere lithography combined with interfacial reactions at the solution surface [61]. Ag nanothorn structures were facilely prepared with the help of thiourea via electrodeposition [62].

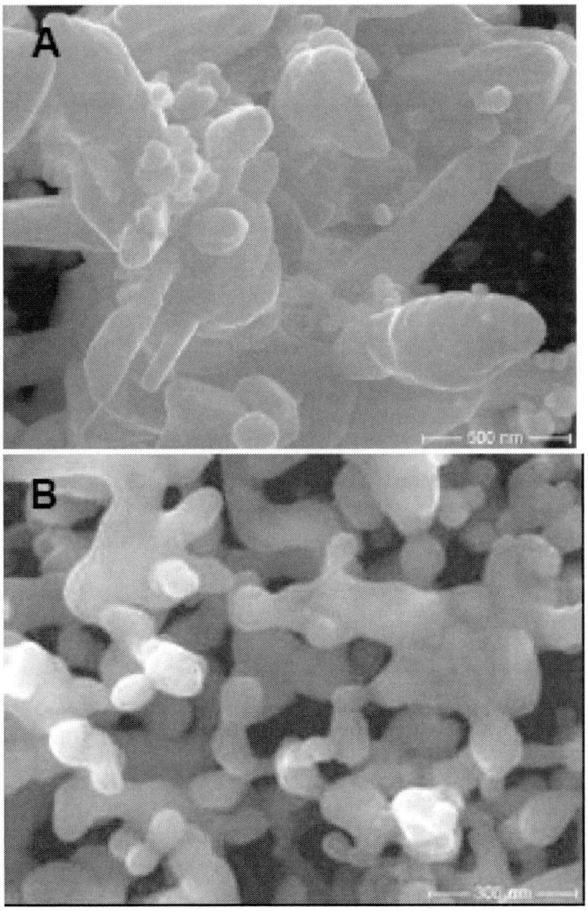

Figure 1. SEM images of Ag electrodeposited onto a graphite plate at -0.5 V from an electrolyte solution containing 5 mM $AgNO_3$ + 0.1 M KNO_3 without (top) and with 0.1 wt.% SDS (bottom).

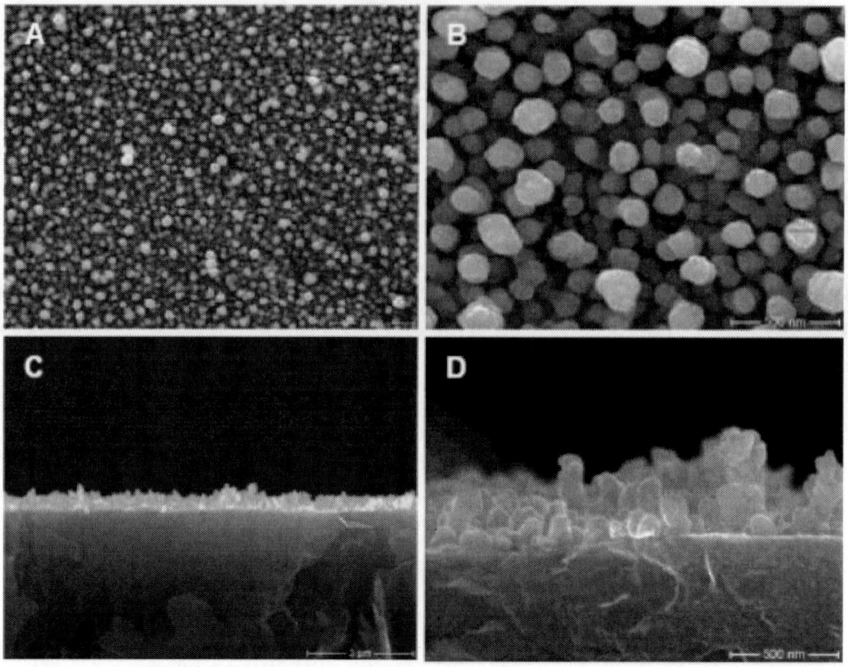

Figure 2. Low (A, C) and high (B, D) magnification of SEM images for Ag nanostructures electrodeposited on a carbon graphite plate at -600 mV for 300 s. Top: top views. Bottom: cross sections.

In general, sodium dodecyl sulfate (SDS) is known to adsorb on electrode surface affecting the kinetics of heterogeneous electron transfer, and thereby direct crystal growth during the electrodeposition process [63-65]. In our group, Ag nanocoral-like structures were potentially grown on the graphite electrodes using SDS as a soft template [66]. In this work, the presence of SDS heavily affects the morphologies of the Ag deposits, in comparison with that without SDS under the identical conditions (Figure 1). The generated nanocoral-like structures are conductive, high surface area to mass ratios, and remarkably stable.

Combining with sufficient NH_4^+ ions, nanopillar-like arrays were formed on the graphite electrode in our lab by simple one-step electrodeposition through diffusion-limited process (Figure 2), using SDS and NH_4^+ ions as structure-directing agents [28]. In this case, Ag^+ ions access to the electrode surface is controlled by diffusion through complexation of Ag^+ and NH_4^+ ions by SDS. Nucleation and lateral/perpendicular growth of the Ag nanopillar structures strongly depends on the deposition potential, which in turn controls the density and diameter of the produced nanopillars (Figure 3). The structures of the prepared pillar-like arrays were further characterized by X-ray diffraction (XRD) spectra (Figure 4), which show the characteristic peaks of crystalline Ag with (111), (200), (220), and (311) facets.

Taking advantage of the similar mechanism, well-dispersed Pt nanoparticles were grown on a glassy carbon electrode [67]. Evidently, various inorganic anions and deposition potential are important to the morphologies and dimensions of the final deposits (Figure 5). The formed Pt particles were further characterized by cyclic voltammetry (Figure 6), where the typical peaks for hydrogen adsorption-desorption and metallic Pt are observed clearly.

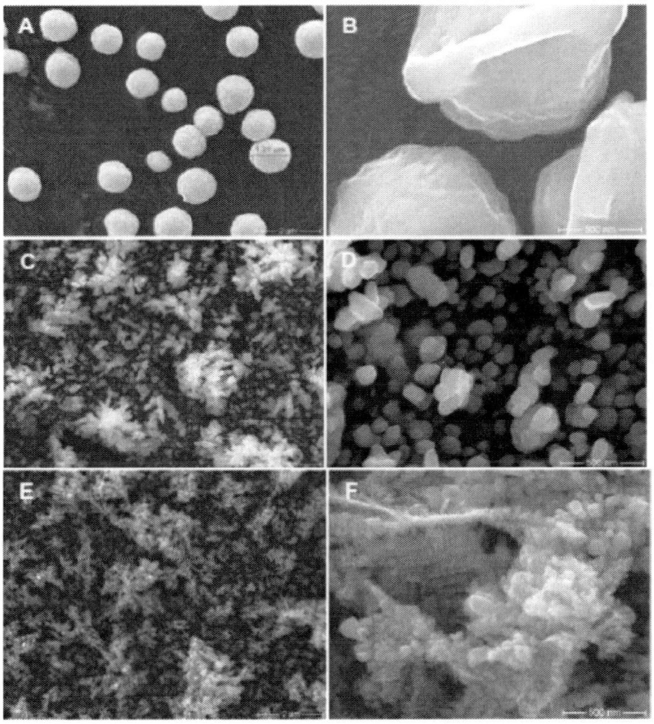

Figure 3. SEM images of Ag nanostructures electrodeposited on a carbon graphite plate at different potentials for 300 s: (A-B) -100 mV, (C-D) -400 mV, (E-F) -900 mV.

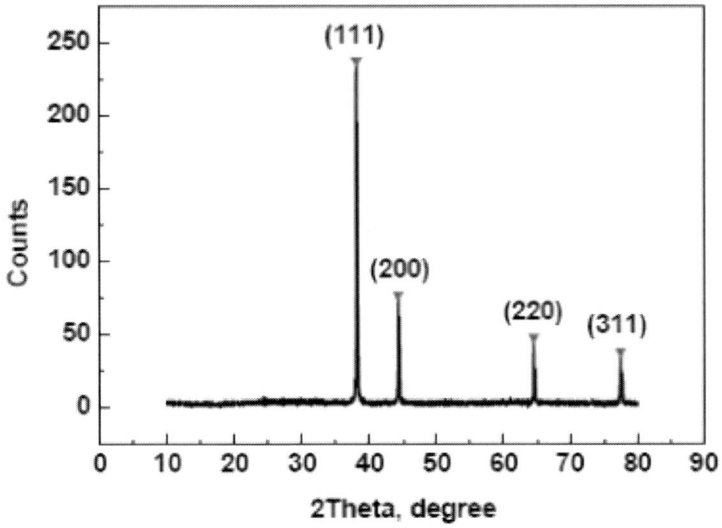

Figure 4. XRD pattern of an Ag nanopillar array prepared as described in Figure 3.

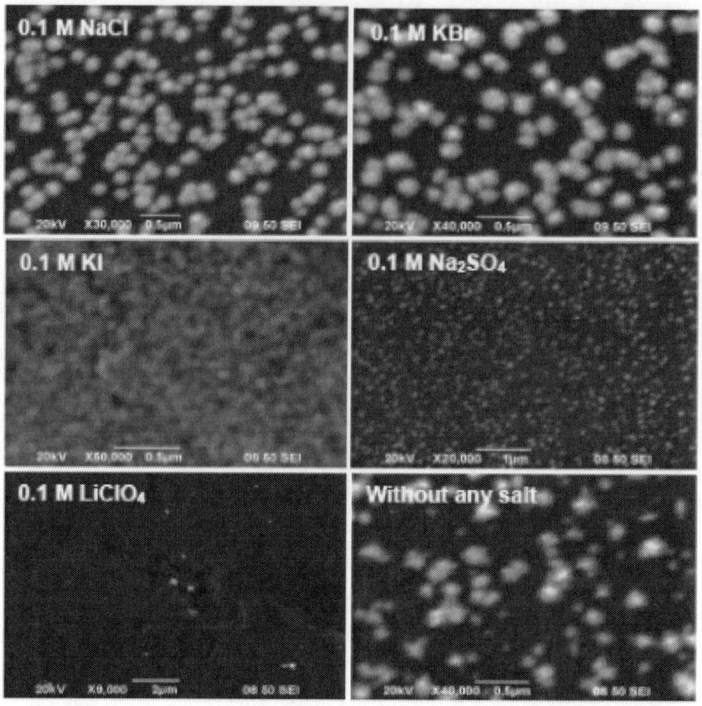

Figure 5. SEM images of the Pt NPs electrodeposited on a GCE at -700 mV for 40 s in 10.0 mM K_2PtCl_6.

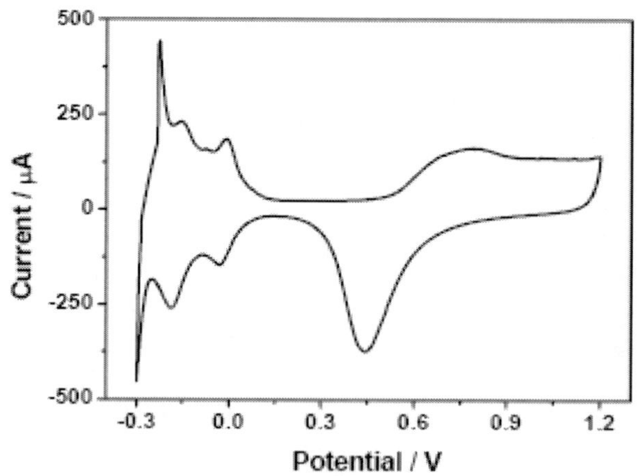

Figure 6. Cyclic voltammogram of the Pt NPs modified electrode in 0.5 M H_2SO_4 at 50 $mV{\cdot}s^{-1}$.

2.2. Noble Metal Based Hybrid Nanostructures

Noble metal substrates containing single Ag or Au, either with SERS-active roughened or functional surfaces, have severely limited their further applications in SERS and biosensing fields. As a result, synthesis of hybrid structures combined the advantages of each noble metal becomes urgent. Up to now, promising progress has been made by many groups [34,68-70], such as Au-Ag based core-shell structures [71], and the direct (electro)deposition of Au (or Ag) hybrid nanostructures [72-76]. Through atomic layer deposition process, Ag coated Pt core–shell nanostructures on etched Si nanowires were constructed and the intensity of the SERS enhancement was discussed based on the finite element modeling [75]. The shape and size of the metal nanoparticles significantly determine the strength of SERS enhancement because these factors affect the ratio of absorption and scattering events. Bimetallic Ag–Au anisotropic nanostructures were assembled by means of electrostatic interactions [77]. Heterogeneous Ag–Cu dendrites were grown from a Cu foil by a galvanic displacement reaction at room temperature [78].

In general, Au (or Ag) based hybrid structures is difficult to keep their individual chemical and physical properties, especially for the core counterparts [79]. For example, upon straightforward deposition of Au on Ag surfaces, the SERS enhancement of the target molecules adsorbed on the outside Au surfaces were dropped down [33,45]. An even stronger decrease of SERS enhancement was observed for Au nanoparticles directly coated with ultra-thin shells of various SERS-inactive transition metals (Pt, Pd, Ni or Co) [80]. Conversely, randomly oriented Ag nanoparticles were coated with a thin dielectric layer to form multilayer substrates, which is observed to tailor both the extinction spectrum and the intensity of SERS enhancement by varying the thickness of the thin dielectric layer [46]. Similarly, particular high SERS enhancement were noticed on the core-satellite geometries that formed by hybridization of linker oligonucleotides to thiolated single-stranded DNA bound to the Au nanoparticle surfaces [81], as well as on the two-dimensional metal nanoparticle arrays under periodic boundary conditions, owing to strong electromagnetic fields between two metal layers [82].

Therefore, various strategies have extended the technique to non-traditional substrates [83-86]. Ag-coated colloidal crystal films were used as SERS-active substrates with very large enhancements [87]. The intensities of SERS enhancement were adjusted with a series of periodic SERS-active Ag nanowires that were embedded in highly ordered porous aluminum oxide, by varying the inter-wire gap distance from 10 to 35 nm [88]. Ag nanoparticles were successfully coated with ultrathin diamond-like carbon films, which efficiently protected the oxygen-free Ag in air and prevented Ag ionizing in aqueous solution, and revealed high SERS enhancement [84]. Shell-isolated nanoparticles were constructed by coating the Au nanoparticles with an ultrathin silica or alumina layer. The ultrathin coating prevents the particles from aggregation, keeps them from direct contact with the probed material, and allows efficient assembly for the formation of different shapes on the substrates [83].

In our group, Ag-Au hybrid multilayered devices were designed and constructed, revealing a long-range plasmon transfer from Ag to Au surface, where SERS enhancement is offered by inner Ag layer while surface reactions occur at the Au surface outside. The device includes a rough nanoscale Ag surface, which is separated from a Au layer by a thin dielectric layer, such as a series of amino-terminated mercaptoalkanes with different chains [29,89] or a thin silica film with tunable thickness between 1 and 21 nm [30].

Self-assembly technique generally provides convenient, flexible, and simple system for tailoring and fictionalization of the interfacial properties of the metal electrodes [4,5,90,91]. The self-assembled monolayers (SAMs) on metal electrodes are widely employed in electrochemical and biochemical analysis [4,5], or nanotechnology [90,92]. In our study, for the construction of Au-SAMs-Ag hybrid devices, SAMs of amino-terminated mercaptoalkanes was introduced as a dielectric layer between the rough Ag substrates and a thin Au layer, as described in the scheme (Figure 7A). As SEM images demonstrated, a porous surface structure with a grain size between 60 to 100 nm is observed on the rough Ag electrode (Figure 7B). The Au islands obtained with an average lateral dimension of 300-400 nm and a thickness of less than 20 nm (Figure 7C-D). Additionally, the whole procedure for the formation of the external Au layer could be monitored by SEM experiments. Obviously, with time evolution, such as 0.5 h (Figure 8A) and 2 h (Figure 8B), the morphologies of the subsequent formed Au nanostructures are different, much distinguished with those formed after 4.0 h (Figure 7C-D), However, the mercaptoalkanes are usually expensive and instable upon exposing to air, which severely limited its further applications. In addition, these strategies with metal electrode can not generally be transferred to other non-metal supports such as glassy carbon electrodes, due to the specific metal-sulfur bonding.

Using silica instead of mercaptoalkanes-based SAMs could substantially improve the long-term stability while reducing the production costs definitely. Furthermore, the long-range optical responses of the hybrid device could be precisely tuned by varying the thickness of the silica layer.

Figure 7. (A) Schematic illustration of the fabrication of the Au-SAMs-Ag hybrid devices. SEM images of SER-active Ag and Au-Ag hybrid electrodes. (B) Ag, top view, (C) Au-Ag, top view, and (D) Au-Ag, scratch through the layer.

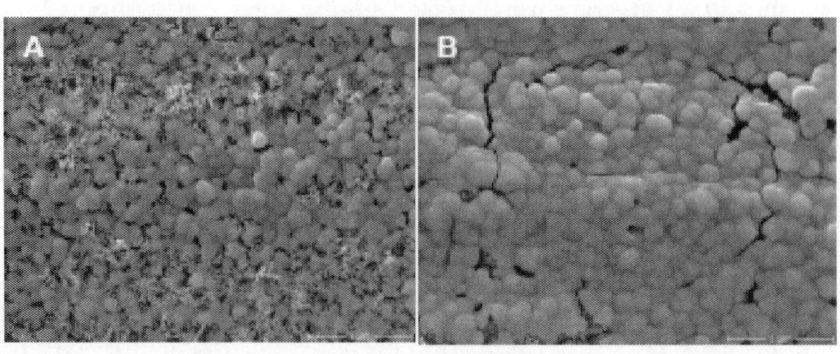

Figure 8. SEM images for monitoring the formation of Au outer layer at different time: (A) 0.5 h and (B) 2.0 h.

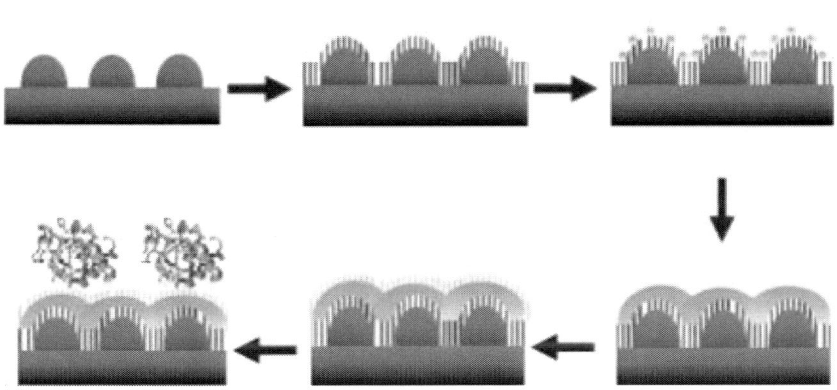

Figure 9. Scheme for the preparation of the Ag–silica–Au devices.

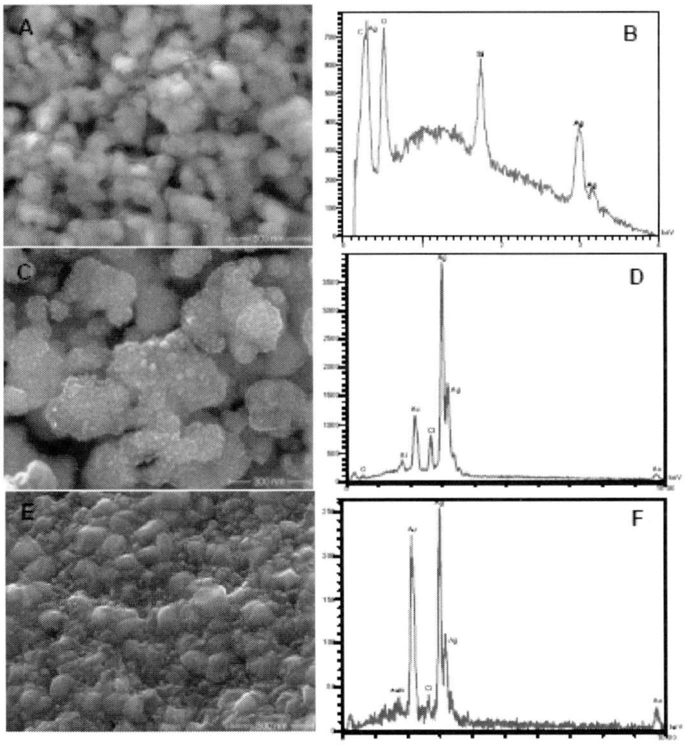

Figure 10. Left panel: SEM images with high magnification of A) Ag-t-SiO_2-a (d_{Si} = 20 nm), C) Ag-t-SiO_2-a-Au seed, and E) Ag-m-SiO_2-a-Au film. Right panel: SEM images with lower magnification of B) Ag-t-SiO_2-a (d_{Si} = 20 nm), D) Ag-t-SiO_2-a-Au seed, and F) Ag-m- SiO_2-a-Au film, along with the corresponding EDX spectra.

Besides, the functional silica layer allows further assemblies with versatile nanoparticles to form different morphologies on the substrates. Interestingly, this layer has the additional ability to improve the optical performance of the device. Finally, an Au island film is formed on top of the silica surface by seed-growth method[10]. The scheme for the preparation of the

Ag–silica–Au biodevices was presented (Figure 9), and each step for the formation of such devices was demonstrated by SEM images (Figure 10).

On the basis of the SEM images and the energy dispersive X-ray spectra (EDX), uniform and transparent silica layer was clearly formed around the Ag nanostructures (Figure 10A-B). After the attachment of Au nanoparticles, many small Au particles were decorated onto the silica surface (Figure 10C-D). Subsequently, the Au particles decorated electrode was treated with the Au plating solution, and consequently a homogeneous Au layer was formed on the top of the silica. As can be seen from the cross-section SEM image, the shape of the Au island film with a thickness of 20 nm under optimal conditions differs from that of the inner rough Ag nanostructures (Figure 10E-F). Additionally, the outer Au surface is further functionalized by a biocompatible layer for electrostatic adsorption of Cyt-c, which is described in more detail afterwards [29,30,45].

3. Construction of Bioelectronics Devices

The specific properties of proteins concerning redox behavior, molecular recognition, and catalysis represent an important potential for the design of bioelectronics devices [93]. Construction of such devices requires immobilization of the target protein with native structures, efficient electronic communication, sufficient long-term stability, efficient catalysis, and a high protein loading. Accordingly, different strategies have been developed for efficient adsorption of proteins onto different substrates, including SAMs [5,14,15,22], nanostructured materials [94-97], conducting polyelectrolyte [98,99], and sol-gel films [100]. Particularly, the metal electrode is further coated by self-assembly of mercaptoalkanes for better biocompatibility and stability, avoid protein denaturation, and allow for reversible and fast redox processes [4,5,19]. Furthermore, the biocompatible coated metal electrode represent simple but instructive biomimetic systems for biological membranes where most of the redox proteins exert their natural functions [1,5,15,22].

3.1. Biocompatible Ag Nanostructures

Generally, direct adsorption of redox heme protein such as Cyt-c onto the rough Ag electrode would inevitably cause severe protein denaturation [15,20], owing to strong electric fields in the vicinity of the immobilized protein. This assumption has been demonstrated by the denaturation procedures with controlled external potentials [21]. Conversely, biocompatible coatings with dielectric materials would change the size and dielectric constant of the surrounding medium close to the metallic surface, and consequently preserve the protein with native structures. To date, quite a lot of biocompatible coatings have been constructed by a series of mercaptoalkanes with different head groups such as $-NH_2$, –COOH, and –OH groups [101-104], for the formation of a stable metal-sulfur bond. As demonstrated, these SAMs provide a good microenvironment for efficient adsorption of the proteins on metallic surfaces [5,15,22]. These mercaptoalkanes are commercially available with versatile alkane chains and head groups. The head group that exposes to the solution can be changed and is usually chosen with regard to the target protein under investigations. As a result, the interactions between the electrode and protein would change by varying the surrounding environment. Furthermore, the protein is isolated from the metal surface by SAMs and thereby reduced the SERS intensity significantly [15,89,92].

Up to now, Ag is the best choice as the SERS-active substrates for investigating the immobilized Cyt-c that shows strong absorption in the Soret region around 410 nm [105,106], which is widely used as a model protein for studying various reaction steps of interfacial redox processes[4,5,14]. Much work has been carried out for studying how the local electric fields affect the structure, dynamics, and function of the immobilized protein. These results have been reviewed elsewhere [4,5,105] and will not be discussed here in detail. A significant emphasis will be focused on the recent work performed by our group [107-110].

Typically, biocompatible coating of the rough Ag electrodes can be obtained by putting the electrode into a solution containing designated ω-functionalized mercaptoalkanes overnight. Different head groups of mercaptoalkanes that face the solution, hybrid SAMs with different ratios, and pH have great effects on the efficient adsorption of the protein, as well as on its dynamic and catalytic performance on the surface [107-110]. Ag nanocorals were grown up at the graphite electrodes and further coated with mercaptohexanoic acid for efficient adsorption of Cyt-c[28]. The scheme for the construction of the nanocoral-based biointerfaces was illustrated (Inset in Figure 11). Evidently, the immobilized protein retained

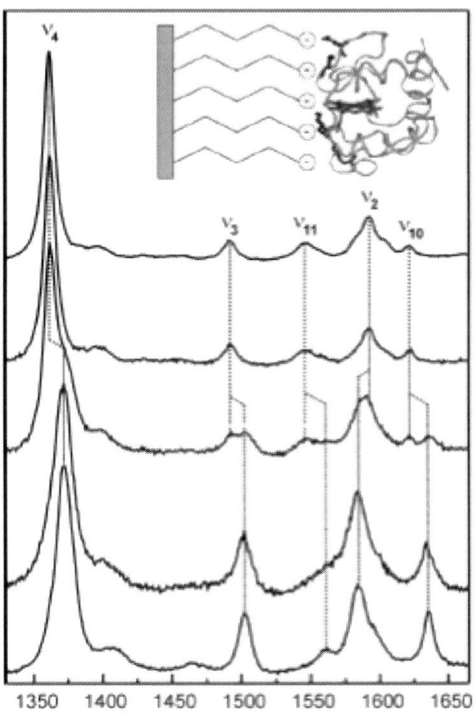

Figure 11. RR and SERR spectra of Cyt-c in solution and adsorbed onto mercaptohexanoic acid coated Ag nanocoral electrodes, respectively. Top to bottom: RR of ferrous Cyt-c; SERR at -400 mV; SERR at 0 V; SERR at 150 mV; and RR of ferric Cyt-c. Measurements were performed with 413 nm excitation. Inset: Scheme of the construction of the Cyt-c/SAMs-Ag nanocoral-based biointerfaces.

its native structures. Specifically, the analysis in respect to the series of SERS spectra confirms that only native ferric and ferrous Cyt-c is presented (Figure 11). In contrast, Cyt-c directly adsorbed on the bare Ag surfaces (i.e., without a SAM coating) shows the typical spectral features of non-native species (data not shown), similar to previous observations [4]. In another example, a one step electrochemical method was developed for the construction of Ag pillar arrays, which were further functionalized with cysteamine by self-assembly. The attached amino groups could subsequently react with hemin by covalent-linkage [31].

Ultra high charge density on the SAMs surface is not good for efficient adsorption of the immobilised protein, and even severely destroys the protein's three-dimentional structures [15,22,92]. Meanwhile, larger dipole moment of the protein is also undesired. This effect is evidently exemplified by the comparison of the Cyt–c from either common horse heart or yeast that shows a single surface cysteine on position 102 (yeast iso-1 cytochrome c, simply referred as YCC)[14]. The former could preserve its native structure and functions on a rough Ag electrode coated with negatively charged 11-mercaptoundecanoic acid (MUA) (Figure 12).

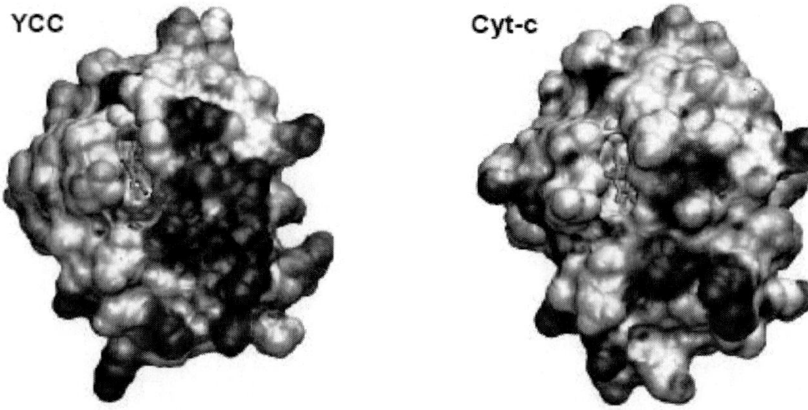

Figure 12. Surface electrostatic potential plots of YCC and Cyt-c. Blue, positive potential; Green, heme centre ; Red, negative potential.

Conversely, similar strategy is not suitable for the case of YCC, where the positive charges are unevenly decorated on the protein surface. YCC is more sensitive to surrounding medium [111], due to larger dipole moment of the protein itself (Figure 12). The electron communication is failed and a slow denaturation process is observed in pH 7.0 phosphate solution [21], while different observations obtained in pH 6.0 phosphate solution. However, the charge density at the SAMs/protein interface can be tuned by changing pH or the composition of the SAMs. In our work, for reducing the charge density at the interface, carboxyl-terminated mercaptoalkanes is mixed with less polar hydroxide-terminated mercaptoalkanes, including 1:1 mixtures of MUA and either 11-mercaptoundecanol (MU) or 7-mercaptoheptanol (MH). In both cases, the immobilized YCC could retain its native structures. Similarly, the protein denaturation is suppressed and the protein remains redox active on the surface in pH 6.0 phosphate solution. In all cases, better performance is observed at pH 6.0 in comparison with that at pH 7.0.

In addition, for negative charged protein, such as cytochrome b562, was electrostatically immobilized on Ag electrodes coated with positively charged SAMs of amino-terminated alkanethiols. The conformational and redox equilibria of the immobilized protein are compared to those of the positively charged Cyt-c immobilized on negatively charged electrode coatings [108].

3.2. Biocompatible Au Nanostructures

The bare Au, similar to other noble metals, has the strong interactions with the target proteins, causing large changes in the three-dimensional structures and often in denaturation of the protein [112-114]. Great efforts have been made towards keep the protein from denaturation upon adsorption through the modification of the electrode surfaces [14,40,115-120]. Upon electrostatic binding of Cyt-c to Au electrodes coated with SAMs of carboxyl-terminated mercaptoalkanes, the immobilized Cyt-c retains its native structures and display a reversible redox process with a redox potential, similar to that of Cyt-c in solution [92]. When Cyt-c is attached to a gold surface that coated with MUA, the deprotonation of the carboxyl head group of the SAMs is essential for subsequent efficient adsorption of the protein. Clearly, the electron transfer of Cyt-c is the fastest only with moderate amount of the carboxyl head group deprotonated [119].

The Au-Ag hybrid device consists of a nanostructured Ag support, which is separated from an Au film by SAMs. For biocompatible immobilization of YCC, the Au surface was previously coated with a hybrid SAMs of MUA and 11-mercapto-1-unodecanol (MU) by exposing the Au-Ag electrode into an equimolar (1 mM) ethanolic solution of MUA and MU for 24 h [29]. This device shows the electrochemical properties of an Au electrode while the optical properties are very similar to that of Ag surfaces, indicating that Ag is completely isolated from aqueous solution by the Au layer outside. The results reveal that the immobilized YCC on the hybrid device remains exclusively native state, unlike those of YCC on roughened Ag electrode using the similar strategy [89,121]. To extending its applications, a series of amino-terminated mercaptoalkanes were employed to form the Ag-(C_x-NH_2)-Au devices (C_x-NH_2, with x = 6, 8, 11). The top Au surface is biocompatibly coated with different chain length of carboxyl-terminated mercaptoalkanes (C_x-COOH) surface to bind YCC as a benchmark protein, similar results were obtained on these hybrid devices.

Lately, a novel Ag–Au hybrid device was developed by our group using silica as an insulating layer between the two metals. Thin silica layers are generated from silane precursors by sol-gel method, which are strongly adhered onto Ag and Au surfaces [122-124] and widely used in technical applications [125-128], owing to their non-toxicity, chemical inert, good stability, and low cost. To study the biocompatibility of the Ag–silica–Au devices, the outer Au surface was functionalized with MUA through self-assembly, and further immobilized Cyt-c by electrostatic interaction. High-quality SERR spectra are obtained with 413 nm excitation by time- and potential- dependent SERR spectroscopy. We notice that this immobilization strategy can preserve the protein with native structures and reversible redox behavior [30].

4. APPLICATIONS

Biofunctionalization is an important step that will assist the metal structures to develop various biointerfaces for detection and identification of small molecules, nucleic acids, lipids, peptides, proteins, as well as *in vivo* cellular sensing [2,9,36,40,51,53,95,97,99,101,129]. Many mecaptoalkanes with different head groups are usually employed for this target. For example, functionalized Au and Ag surfaces are used for efficient protein immobilization, interfacial electron communication, and biocatalysis. In respect to these investigations, most common techniques such as SERR spectroscopic and electrochemistry, are widely used for detection of proteins [43,117,130], DNA [13,27,131], peptides [58], cells [61,66,86,100,118,131-134], and so on.

4.1. Biosensing Devices

Membrane proteins perform most of the important processes at external and internal membrane of all living cells. In general, the local electric fields strongly affects the structures, the reaction mechanism and dynamics of the protein such as Cyt-c [3,5,15,107,110,130]. To investigate the electric field effects, a variety of biomimetic interfaces were prepared, which contain the target protein bound on biocompatible and functionalized metal substrates. A series of reaction steps of interfacial redox processes have been systemically investigated in respective to rotational diffusion, electron transfer, as well as protein and cofactor structural changes [4,5,22,92,119].

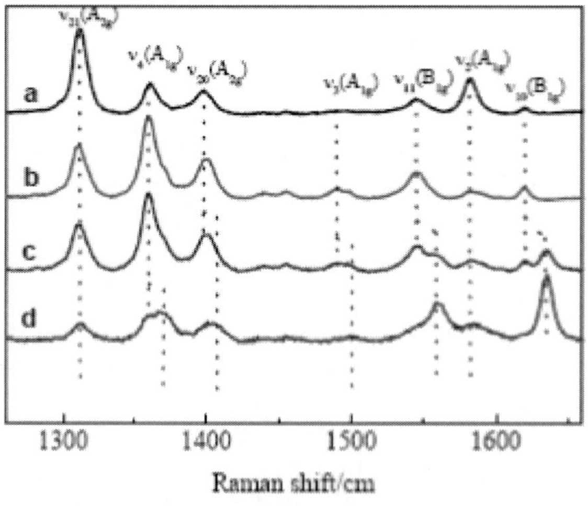

Figure 13. RR and SERR spectra of YCC adsorbed on MUA/MH in pH 7.0 phosphate solution. From top to bottom: RR of (black) ferrous YCC; SERR of ferrous YCC at (olive) -400 mV; SERR at (blue) 50 mV; and SERR at (wein) 120 mV; RR of (red) ferric YCC. Excitation laser: Ar 514nm.

4.1.1. SERS-Active Substrates for Heme Proteins

For YCC immobilized on MUA and MUA/MU coated Ag rough electrode at pH 7.0 and 6.0 phosphate solution[111], SERR spectroscopic measurements were conducted with 413 nm excitation that allows direct optical excitation of surface plasmons of Ag, and exclusively display the vibrational bands of the heme cofactor since this wavelength is in resonance with its strong electronic Soret transition [3-5,22,89]. In all cases, the immobilized YCC showed high quality SERR spectra, which allows for *in situ* structural characterization of the immobilized proteins (Figure 13). The redox potentials were negative shifted in comparison with that of the protein in solution. In each system, the rate constant for the protein reorientation was identical to the electron transfer (Table 1).

Table 1. Thermodynamic and kinetic parameters of YCC adsorbed on SAM-coated Ag electrodes[a]

SAMs	pH	E_{pzc}	E°/ mV	□E_{pzc}°/ mV	E_{RC}/ mV	k_{orient}/ s^{-1}	K_{redox}/ s^{-1}	I(□10/□4)[b]
MUA	6.0	-350	34	384	-48	5.0	5.2	0.2
	7.0	-450	16	466	-66	n.d.	n.d.	0.8
MUA/MU	6.0	-350	45	395	-37	8.1	7.5	0.6
	7.0	-380	30	410	-52	5.0	4.8	0.6
MUA/MH	6.0	-280	29	309	-52	17.8	18.0	1.5
	7.0	-350	24	374	-57	9.3	8.6	1.2
solution	7.0		82					5.6

E_{PZC}: Potential of zero charge derived from capacitance measurements. $E°$: Redox potential versus Ag/AgCl. $\Delta E_{PZC}°$: Difference between $E°$ and E_{PZC}. E_{RC}: Redox shift $|E° - E_s|$. k_{orient}: Rate constant for protein reorientation for jumps from E_i) +80 mV to E_f = -20 mV. k_{redox}: Apparent redox rate constant for jumps to $E_f = E°$. b Intensity ratio ν_{10}/ν_4 measured at 50 mV.

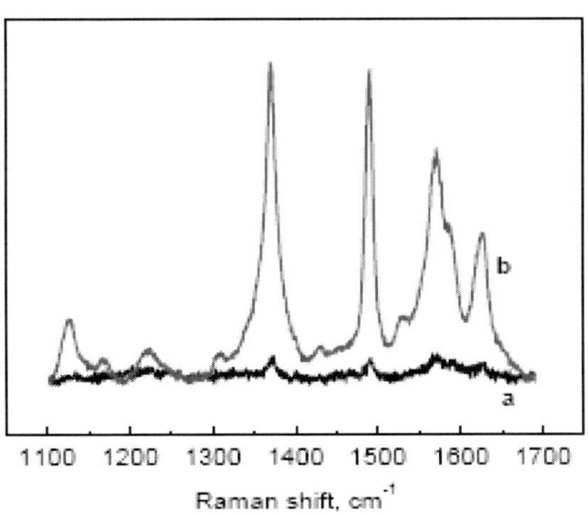

Figure 14. (A) RR spectrum of hemin in solution (50 μM) and (B) SERR spectrum of hemin covalently attached to an Ag nanopillar electrode. Both measurements were performed under otherwise identical conditions.

The results indicate that the protein reorientation is the rate limiting step for the whole interfacial redox process, distinctly slower than that for typical Cyt-c under similar conditions [3].

In the same way, Cyt-c adsorbed onto mercaptohexanoic acid coated Ag nanocoral structures with external potentials sufficiently negative or positive, the redox potential was estimated from the SERR spectra and found identical to the resonance Raman spectra of ferrous and ferric Cyt-c in solution, respectively [66]. In another example, when hemin was cross-linked onto the cysteamine functionalized pillar-like Ag arrays [31], which provide reproducible SERR signals within 6% standard deviation. The surface enhancement factor of the arrays electrodeposited at -500 mV was determined to be about 1 × 10^5 (Figure 14), by comparison of SERR spectra of the covalently attached hemin and RR spectra of the same compound in solution [135].

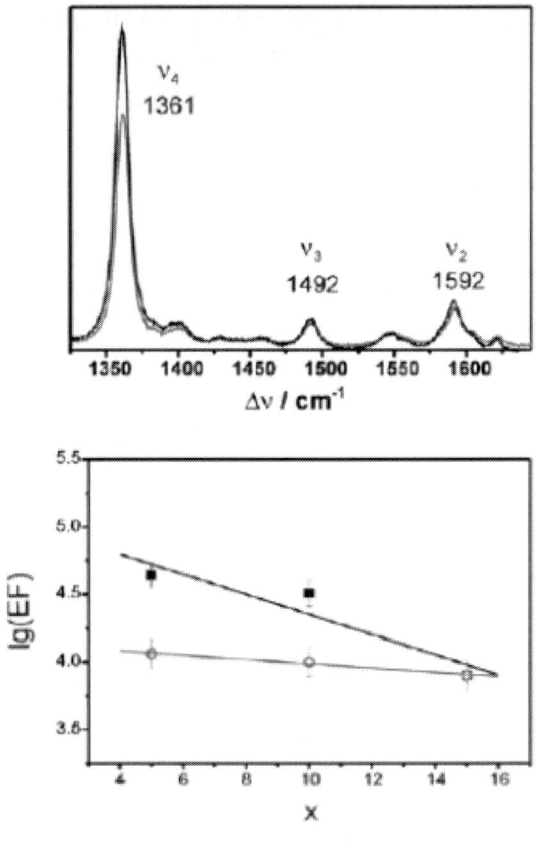

Figure 15. (A) SERR spectra of YCC on Ag (black) and Ag-(C_x-NH_2)-Au (grey) electrodes at E = -0.4 V with 413 nm excitation. The spectra were obtained with the same laser power and accumulation time. (B) Semi-logarithmic plot of the enhancement factors for Ag-(C_x-COOH) (hollow circles) and Ag-(C_{11}-NH_2)-Au-(C_x-COOH) (solid squares) as a function of methylene groups of the carboxylterminated SAM (x = 5, 10, 15). The error bars indicate the variation for measurements from different preparations.

Nanostructured Au-SAMs-Ag devices advantageously combine the electrochemical properties and chemical stability of Au and the strong SERS enhancement by Ag. Remarkable SERS enhancement was observed for YCC immobilized on this hybrid devices, in comparison with that of YCC immobilized on SAMs coated Ag or using (poly)ethyleneimine layer as a spacer instead of amino-terminated mercaptoalkanes [29]. This unexpected enhancement was ascribed to the optical excitation of the Ag surface plasmons that are coupled to the Au surface plasmons, leading to an enhancement of the resonance Raman scattering of the nearby heme group. Therefore, an insulating thin film is required for optimal surface enhancement via the Au surface, supported by those previously published theoretical and experimental results [33,83,136,137]. The devices were further explored for analyzing interfacial electron transfer processes of the YCC by stationary and time-resolved SERRS (Figure 15) [89].

By electrostatic adsorption of Cyt-c, the protein exclusively remained in its native structure, and demonstrated efficient electron communication with the electrode (e.g. the apparent number of transferred electrons (n) is 0.8-0.9). That is, a fast tunneling electron transfer was observed across the long-distance multilayered Ag-SAMs-Au-SAMs nanostructures [89]. Conversely, the covalent attached YCC through the metal-sulfur bond exhibited two broad redox transitions (n = 0.3), owing to part of them transforms to non-native species that remained redox inactive in the potential window between 0.1 V and -0.3 V.

In a similar way, Ag–silica–Au hybrid device also displays a long-range plasmon transfer from Ag to Au, as confirmed by Cyt-c efficient adsorption onto the outer MUA-coated Au surface, not the optically excited Ag surface [30]. The SERS enhancement exclusively come from excitation of Ag surface plasmons with an enhancement factor of 2×10^4 – 8×10^3 for a separation of Cyt-c from the Ag surface by 28–47 nm. This value corresponds to an attenuation of the enhancement by a factor of only 2–6, in comparison with that of Cyt-c directly adsorbed on a SAMs coated rough Ag electrode. The unusual long distance enhancement is due to the dielectric silica layer, which separates the optically active Ag support and the outer Au film, leading to an effective long-range plasmon excitation transfer from Ag to Au. Furthermore, the intensities of the enhancement of Cyt-c on Ag–silica–Au–MUA devices can be tuned by varying the thickness of the silica layer (Figure 16). Additionally, the intensity justly moderately decreases by a factor of 2.5 with the increase of silica thickness from 1 to 20 nm. Even more distinguished is that the SERRS intensity for such devices is only 2–6 times lower, in comparison with that without any silica layer. The redox potential and the apparent number of transferred electrons (n) were estimated according to the Nernst equation (Table 2). These plots display a good linear correlation, while the values for n decreased with the increase of silica thickness, suggesting a more heterogeneous adsorption behavior of the protein ensemble. The redox potential of the adsorbed Cyt-c shows a slight negative shift (54 mV at pH 7.0), compared to the Cyt-c in solution. The redox potentials decrease with the silica thickness, similar to our previous findings for Cyt-c bound to Ag electrodes coated with different SAMs [29]. The reason is ascribed to the potential drop across the SAMs that increased with the increase of the layer thickness. In addition, an impressive reusability of the device was observed through repeated removal and re-binding of MUA and Cyt-c without substantial loss of the optical performance (Figure 17).

Table 2. Thermodynamic and kinetic parameters of the heterogeneous electron-transfer process of the Cyt c on Ag–silica–Au–MUA electrodes

d_{Si}/nm	E^0/mV	n	k_{app} / s^{-1}
1.0	2.4	0.78	2.3
2.5	0.9	0.7	1.8
11.0	-10.0	0.61	1.3
21.0	-25.1	0.57	1.1

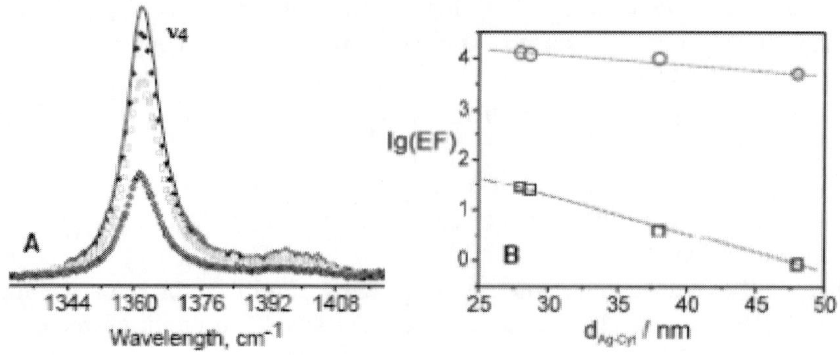

Figure 16. (A) SERR spectra of Cyt-c (v_4, 1361 cm^{-1}) adsorbed on various MUA-coated Ag–silica–Au devices at -0.4 V: Ag-a-SiO_2-Au (1 nm, solid), Ag-m-SiO_2-a-Au (1.7 nm, solid circles), Ag-t-SiO_2-a-Au (10 nm, hollow squares), Ag-t-SiO_2-a-Au (20 nm, hollow circles) electrodes coated with MUA-SAMs. B) Enhancement factors for Cyt-c immobilized on MUA-coated Ag–silica–Au devices as a function of the distance of the protein from the Ag surface (hollow circles). The hollow squares indicate the values predicted according to Equation (1). The dotted lines are just to guide the eyes.

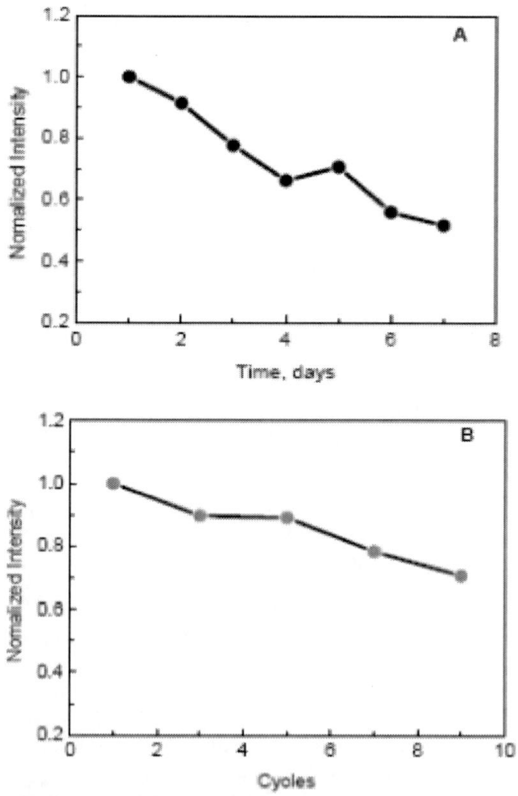

Figure 17. Reusability of the Ag-m-SiO_2-a-Au devices (A, black circuits) SERR intensity of Cyt-c following protein desorption and re-adsorption cycles. (B, grey circuits) SERR intensity of Cyt-c following protein and MUA SAMs desorption and readsorption cycles.

4.1.2. Electrochemical Biosensors

The SERS-active conducting substrates can also work as working electrodes by applying an external electric circuit in an electrochemical cell. The coral-like Ag nanostructures provides a suitable platform for protein-based bioelectronics devices [28]. The adsorbed Cyt-c shows direct electron transfer of the protein with the electrode by cyclic voltammetry (Figure 18).

The modified electrodes were exemplified by the fabrication of a hydrogen peroxide amperometric sensor based on Cyt-c with broader linear range up to ca. 23 mM (Figure 19), compared to that of Cyt-c adsorbed onto three-dimensionally ordered macroporous active carbon [138] and poly(ferrocenylsilane)–DNA composite film modified gold electrode [139]. Meanwhile, the hemin covalently coated functional pillar-like arrays display fast electron transfer with the electrode and good mediatorless electrocatalytic activity toward the reduction of NO_2^- ions (Figure 20), and thereby presenting good prospects for its applications in sensing devices.

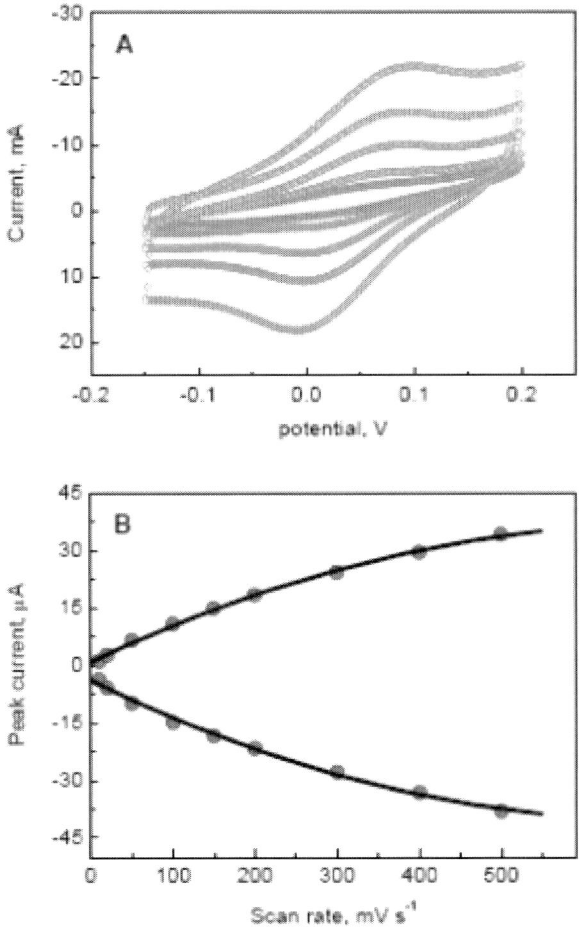

Figure 18. (A) Cyclic voltammograms of Cyt-c adsorbed on SAMs-coated Ag nanocoral electrodes recorded at different scan rates (10, 20, 50, 100, and 200 mV s^{-1}) in 10 mM PBS buffer at pH 7.0; (B) peak currents as a function of the scan rate.

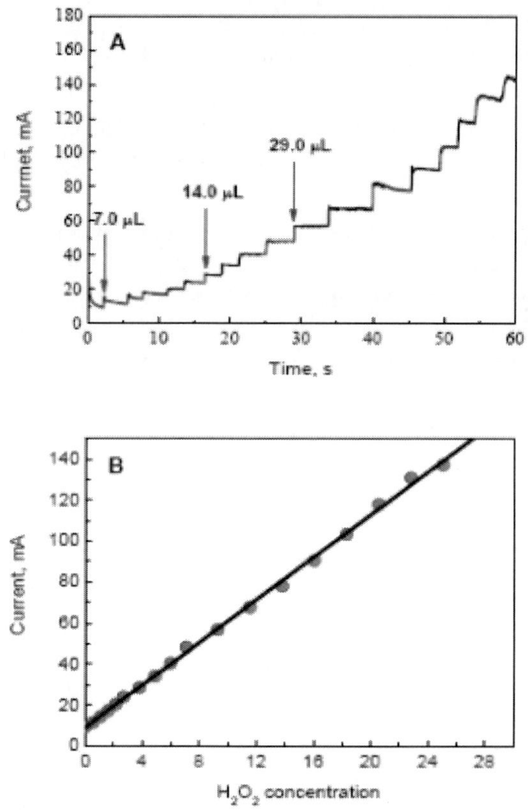

Figure 19. (A) Amperometric responses of graphite/nanocoral/SAMs/Cyt-c electrodes at -100 mV upon successive additions of different volumes of 5 mM H_2O_2 to the electrochemical cell containing 5 mL of 10 mM phosphate solution at pH 7.0; (B) linear response of the electrocatalytic current with H_2O_2 concentration.

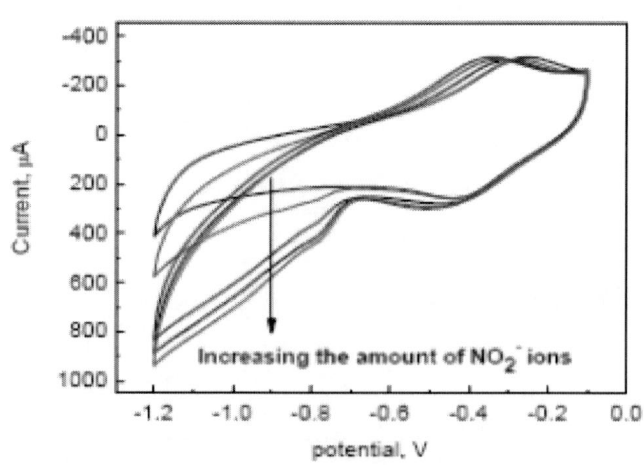

Figure 20. Cyclic voltammograms of a hemin-coated nanopillar electrode array recorded in the presence of different concentration of NO_2^- ions.

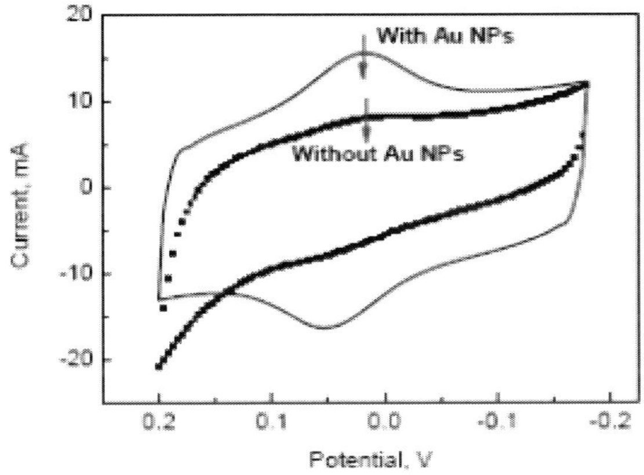

Figure 21. Cyclic voltammograms of Cyt-c adsorbed on Ag (black) and Au-Ag (gray) electrodes coated with MUA/MU-SAMs (scan rate: 0.1 V/s).

The cyclic voltammograms (CVs) of YCC immobilized on the coated Au-SAMs-Ag device shows nearly symmetric oxidative and reductive peaks, implying a reversible redox reaction and efficient electron communication of the protein with the metal substrate (Figure 21). The redox peaks of YCC are distinctly stronger than those from the SAMs-coated Ag electrodes under the same conditions. Specifically, the surface coverage of YCC was estimated to be ca. 8.5×10^{-11} mol·cm^{-2}, much higher than that of YCC immobilized on SAMs-coated Ag electrodes with a value of 2.3×10^{-11} mol·cm^{-2}. Furthermore, nearly no oxidative peak of Ag was observed at ca. 0.15 V in this system. This finding indicates that the Au-Ag hybrid device displays the electrochemical properties of gold with wide potential window, but not Ag electrode, implying that Ag is totally isolated from contact with the aqueous solution. Moreover, the multilayered Au-SAMs-Ag devices were further employed to study the interfacial electron transfer processes. The redox potentials of YCC are slightly shifted with a value of -56 mV, compared with the YCC in solution, more pronounced than that of YCC adsorbed on either MUA coated Ag or Au with a value of -40 mV and -20 mV, respectively (Table 3) [21,121,140]. These shifts are attributed to the interfacial potential drop across the SAMs and can be rationalized based on simple electrostatic considerations.

Table 3. Redox potentials (E^0) and number of transferred electrons (n) for the redox process of cytochrome c (horse heart) on various coated Ag, Au, and Ag–Au hybrid electrodes[a]

	SH-C_6-COOH		SH-C_6-COOH		SH-C_6-COOH	
	E^0	n	E^0	n	E^0	n
Ag no	0.020	0.8	0.020	0.85	0.006	0.65
Ag-Au C_6-NH_2	n.d.	n.d.	-0.001	0.89	n.d.	n.d.
C_8-NH_2	n.d.	n.d.	0.007	0.92	n.d.	n.d.
C_{11}-NH_2	-0.005	0.9	0.004	0.78	0.015	0.88
Au[a] no	0.037	0.81	0.040	0.77	0.030	0.84

4.2. Detection of DNA

Conventionally, fluorescent labeling is generally employed for the detection of non-fluorescent biomolecules, such as DNA and fluorescent biomarkers [141,142]. Recently, tremendous advances have been made toward the development of alternative nonfluorescent DNA detections [42,143-145]. Particularly, label-free optical sensing strategies have begun to emerge as powerful methods, due to the lower cost and higher sensitivity [27]. Using self-assembled Ag nanoparticles decorated Au nanotip arrays, the adsorbed adenine, thymine, cytosine and guanine molecules has been detected at the sub-picomole level in SERS experiments [146]. Through a controlled assembly process of the DNA functionalized nanoparticles as a probe, specific DNA sequences relating to disease were facilely determined by a biomolecular recognition event to 'turn on' a SERS response. This strategy would lead to a new range of diagnostic clinical probes based on SERS detection [147].

Among them, DNA hybridization detections have wide applications in molecular biology and forensic testing [148]. For example, by substituting adenine with 2-aminopurine in the probe DNA sequence, its adenine-containing complementary target DNA was optically detected via its SERS signal [27]. Combining the SERS signal following sequence specific hybridization and enzyme hydrolysis, a novel method of DNA detection is developed with high sensitivity and specificity [149]. This technique can be employed for the detection of unlabelled duplex DNA and is the first step towards a SERS based signal amplification approach.

4.3. Detection of Cells

Fast microbial detection and identification with high sensitivity and selectivity is a great challenging in many fields [36,37,100,150,151], including clinical diagnosis, pharmaceutical, and food processing technology. Recently, many novel microscopic methods were proposed using vibrational spectroscopy to monitor cell biological processes of live or fixed cells without any dye, stains, or other contrast agent [132,133,152,153]. Specifically, the detection of Raman signals inside a live cell is one of the basic goals for investigation of modern cellular biophysics and an important prerequisite for progress in both molecular and nanomedicine. Therefore, a variety of SERS reporter complexes, mainly Au based nanostructures, such as Au nanorods [118], Au nanoparticles [151], Au nanoflowers [154], Au nanoparticle-decorated carbon nanotubes [155], Au–Ag core–shell particles [43], and Au quantum dots [156], have been developed and tested for cell and tissue labeling [37,41].

For example, a SERS-enabled endoscope was obtained via assembly of Au nanoparticles functionalized carbon nanotubes onto the tips of pulled glass capillaries. The endoscope make it possible to penetrate the cell membrane for intracellular probing and remain positioned inside during lengthy SERS measurements without causing damage to the cell [155]. Thus, such devices could extend their applications to biological sensors for real-time *in situ* cellular diagnostics, biological detection, and pharmaceutical research.

In another example, *in situ* spectroelectrochemical technique was employed for the study of electrocatalytic microbial biofilms direct grown on Ag electrodes by SERRS in combination with cyclic voltammetry [134]. This experimental approach provides unexpected structural information about the outer membrane cytochromes involved in the direct electron transfer across the bacterium–electrode interface.

Conclusions and Outlook

During the past few years, significant progress has been devoted to the development of surface enhanced resonance Raman spectroscopy (SERRS)-based biosensing devices, which is good to deeply understand the molecules processes involved in biosensors, bioreactors, and biofuel cells. These processes consist of inter alia molecules recognition, electron communication of the target proteins, and catalytic reaction steps.

Recent investigations demonstrate that, essential for a SERRS device is the generation of reproducible, uniform, inexpensive, biocompatible, and reusable SERS-active substrates for its routine and practical applications. Since SERS enhancements heavily depend on the size, shape, composition, dielectric environment of the substrates. Until now, some initial results based on plasmonically active nanostructures have been obtained with the following advantages:

1) reproducible, uniform, inexpensive and reusable;
2) direct generation of the Raman-active nanostructures on conducting supports;
3) fabrication of the metallic and dielectric nanostructures in a multilayered way;
4) combining the appropriate Raman-active substrates with functionalized technology offers biocompatible interfaces for efficient adsorption of the proteins while at the same time provides sufficient signal enhancement for spectroscopic analysis

The intensity of SERS enhancement can be controlled by varying the optical amplification parameters and the surrounding environment in these systems. In respect to their applications in bioelectronics devices, biocompatible coating is the prerequisite for preserving the native structures of the target protein. For investigation of biological surface reactions, spectroelectrochemical technique is the best choice to study the conformation change, redox behavior, and heterogeneous electron transfer of the immobilized heme proteins as a model. The resulting devices have good performance in biosensors, DNA and cell detection.

Acknowledgments

Financial support by the National Science Foundation of China (NSFC Nos. 20905021; 20805011, 21175118) are gratefully acknowledged.

References

[1] Hartland, G. V. *Chem. Rev.* 2011, *111*, 3858.
[2] Gao, S.; Koshizaki, N. *Anal. Bioanal. Chem.* 2011, *399*, 91.

[3] Murgida, D. H.; Hildebrandt, P. *J. Am. Chem. Soc.* 2001, *123*, 4062.
[4] Murgida, D. H.; Hildebrandt, P. *Accounts. Chem. Res.* 2004, *37*, 854.
[5] Murgida, D. H.; Hildebrandt, P. *Chem. Soc. Rev.* 2008, *37*, 937.
[6] Xia, Y.; Xiong, Y.; Lim, B.; Skrabalak, S. E. *Angew. Chem. Int. Edit.* 2009, *48*, 60.
[7] Xia, Y.; Lim, B. *Nature* 2010, *467*, 923.
[8] Wiley, B. J.; Qin, D.; Xia, Y. *ACS Nano* 2010, *4*, 3554.
[9] Kim, H.; Kosuda, K. M.; Van Duyne, R. P.; Stair, P. C. *Chem. Soc. Rev.* 2010, *39*, 4820.
[10] Lal, S.; Grady, N. K.; Kundu, J.; Levin, C. S.; Lassiter, J. B.; Halas, N. J. *Chem. Soc. Rev.* 2008, *37*, 898.
[11] Mulvihill, M. J.; Ling, X. Y.; Henzie, J.; Yang, P. *J. Am. Chem. Soc.* 2009, *132*, 268.
[12] Liu, Y.; Offenhäusser, A.; Mayer, D. *Biosens. Bioelectron.* 2010, *25*, 1173.
[13] Tel-Vered, R.; Willner, B.; Willner, I. *Biohybrid Electrochemical Devices*; ed.; John Wiley and Sons, Inc., 2010, pp 333.
[14] El Kasmi, A.; Wallace, J. M.; Bowden, E. F.; Binet, S. M.; Linderman, R. J. *J. Am. Chem. Soc.* 1998, *120*, 225.
[15] Khoa Ly, H.; Wisitruangsakul, N.; Sezer, M.; Feng, J.-J.; Kranich, A.; Weidinger, I. M.; Zebger, I.; Murgida, D. H.; Hildebrandt, P. J. Electroanal. Chem. 2011, ASAP.
[16] Willner, I.; Tel-Vered, R.; Willner, B. In *Engineering the Bioelectronic Interface: Applications to Analyte Biosensing and Protein Detection*; Roy. Soc. Chem., 2009; pp 56.
[17] Fleischmann, M.; Hendra, P. J.; McQuillan, A. J. *Chem. Phys. Lett.* 1974, *26*, 163.
[18] Jeanmaire, D. L.; Van Duyne, R. P. *J. Electroanal. Chem. Interf. Electrochem.* 1977, *84*, 1.
[19] Kudelski, A. *J. Raman. Spectrosc.* 2003, *34*, 853.
[20] Hildebrandt, P.; Stockburger, M. *J. Phys. Chem.* 1986, *90*, 6017.
[21] Murgida, D. H.; Hildebrandt, P. *J. Phys. Chem. B* 2001, *105*, 1578.
[22] Khoa Ly, H.; Sezer, M.; Wisitruangsakul, N.; Feng, J.-J.; Kranich, A.; Millo, D.; Weidinger, I. M.; Zebger, I.; Murgida, D. H.; Hildebrandt, P. *FEBS J.* 2011, *278*, 1382.
[23] Choi, S. H.; Kim, Y. L.; Byun, K. M. *Opt. Express* 2011, *19*, 458.
[24] Choi, D.; Choi, Y.; Hong, S.; Kang, T.; Lee, L. P. *Small* 2010, *6*, 1741.
[25] Huang, F. M.; Wilding, D.; Speed, J. D.; Russell, A. E.; Bartlett, P. N.; Baumberg, J. J. *Nano Lett.* 2011, *11*, 1221.
[26] Liu, G. L.; Lee, L. P. *Appl. Phys. Lett.* 2005, *87*, 074101.
[27] Barhoumi, A.; Halas, N. J. *J. Am. Chem. Soc.* 2010, *132*, 12792.
[28] Feng, J.-J.; Hildebrandt, P.; Murgida, D. H. *Langmuir* 2008, *24*, 1583.
[29] Feng, J.-J.; Gernert, U.; Sezer, M.; Kuhlmann, U.; Murgida, D. H.; David, C.; Richter, M.; Knorr, A.; Hildebrandt, P.; Weidinger, I. M. *Nano Lett.* 2009, *9*, 298.
[30] Feng, J.-J.; Gernert, U.; Hildebrandt, P.; Weidinger, I. M. *Adv. Funct. Mater.* 2010, *20*, 1954.
[31] Feng, J.-J.; Lu, Y.-H.; Gernert, U.; Hildebrandt, P.; Murgida, D. H. *J. Phys. Chem. C* 2010, *114*, 7280.
[32] Costi, R.; Saunders, A. E.; Banin, U. *Angew. Chem. Int. Edit.* 2010, *49*, 4878.
[33] David, C.; Richter, M.; Knorr, A.; Weidinger, I. M.; Hildebrandt, P. *J. Chem. Phys.* 2010, *132*, 024712.
[34] Guo, S.; Dong, S.; Wang, E. *Chem-Eur. J.* 2009, *15*, 2416.

[35] Sun, Y. *Adv. Funct. Mater.* 2010, *20*, 3646.
[36] Abbas, A.; Linman, M. J.; Cheng, Q. *Biosens. Bioelectron.* 2011, *26*, 1815.
[37] Abdulhalim, I.; Zourob, M.; Lakhtakia, A. *Electromagnetics.* 2008, *28*, 214.
[38] Akimov, Y. A.; Chu, H. S. *Phys. Rev. B* 2011, *83*, 165412.
[39] Homola, J. In *Frontiers in Planar Lightwave Circuit Technology*; Janz, S.; Ctyroky, J.; Tanev, S., Eds.; Springer Netherlands, 2006; Vol. 216; pp 101.
[40] Hu, M.; Chen, J.; Li, Z.-Y.; Au, L.; Hartland, G. V.; Li, X.; Marquez, M.; Xia, Y. *Chem. Soc. Rev.* 2006, *35*, 1084.
[41] Shen, A.; Chen, L.; Xie, W.; Hu, J.; Zeng, A.; Richards, R.; Hu, J. *Adv. Funct. Mater.* 2010, *20*, 969.
[42] Ravindranath, S. P.; Wang, Y.; Irudayaraj, J. *Sens. Actuators B: Chem.* 2011, *152*, 183.
[43] Luo, Z.; Chen, K.; Lu, D.; Han, H.; Zou, M. *Microchim. Acta* 2011, *173*, 149.
[44] Muniz-Miranda, M.; Del Rosso, T.; Giorgetti, E.; Margheri, G.; Ghini, G.; Cicchi, S. *Anal. Bioanal. Chem.* 2011, *400*, 361.
[45] Feng, J.-J.; Gernert, U.; Sezer, M.; Kuhlmann, U.; Murgida, D.; David, C.; Richter, M.; Knorr, A.; Hildebrandt, P.; Weidinger, I. *Nano Lett.* 2009, *9*, 298.
[46] Shoute, L. C. T. *Chemphyschem* 2010, *11*, 2539.
[47] Kelly, K. L.; Coronado, E.; Zhao, L. L.; Schatz, G. C. *J. Phys. Chem. B* 2002, *107*, 668.
[48] Zhou, Q.; He, Y.; Abell, J.; Zhang, Z.; Zhao, Y. *Chem. Commun.* 2011, *47*, 4466.
[49] Anker, J. N.; Hall, W. P.; Lyandres, O.; Shah, N. C.; Zhao, J.; Van Duyne, R. P. *Nat. Mater.* 2008, *7*, 442.
[50] Li, Z.; Bao, K.; Fang, Y.; Guan, Z.; Halas, N. J.; Nordlander, P.; Xu, H. *Phys. Rev. B* 2010, *82*, 241402.
[51] Kneipp, J.; Kneipp, H.; Kneipp, K. *Chem. Soc. Rev.* 2008, *37*, 1052.
[52] Han, X. X.; Chen, L.; Ji, W.; Xie, Y.; Zhao, B.; Ozaki, Y. *Small* 2011, *7*, 316.
[53] Qian, X. M.; Nie, S. M. *Chem. Soc. Rev.* 2008, *37*, 912.
[54] Wu, D.-Y.; Li, J.-F.; Ren, B.; Tian, Z.-Q. *Chem. Soc. Rev.* 2008, *37*, 1025.
[55] Baek, Y.-K.; Yoo, S. M.; Kang, T.; Jeon, H.-J.; Kim, K.; Lee, J.-S.; Lee, S. Y.; Kim, B.; Jung, H.-T. *Adv. Funct. Mater.* 2010, *20*, 4273.
[56] Li, M.; Yu, X.-F.; Liang, S.; Peng, X.-N.; Yang, Z.-J.; Wang, Y.-L.; Wang, Q.-Q. *Adv. Funct. Mater.* 2011, *21*, 1788-1794.
[57] Ahmad, M.; Yingying, S.; Nisar, A.; Sun, H.; Shen, W.; Wei, M.; Zhu, J. *J. Mater. Chem.* 2011, *21*, 7723.
[58] Su, Y.; He, Q.; Yan, X.; Fei, J.; Cui, Y.; Li, J. *Chem-Eur. J.* 2011, *17*, 3370.
[59] Xiao, J.; Qi, L. *Nanoscale* 2011, *3*, 1383.
[60] Joseph, V.; Schulte, F.; Rooch, H.; Feldmann, I.; Dorfel, I.; Osterle, W.; Panne, U.; Kneipp, J. *Chem. Commun.* 2011, *47*, 3236.
[61] Hong, G.; Li, C.; Limin, Q. *Adv. Funct. Mater.* 2010, *20*, 3774.
[62] Fang, Y.-M.; Lin, Z.-B.; Zeng, Y.-M.; Chen, W.-K.; Chen, G.-N.; Sun, J.-J.; Ren, B.; Tian, Z.-Q. *Chem-Eur. J.* 2010, *16*, 6766.
[63] Choi, K.-S.; Lichtenegger, H. C.; Stucky, G. D.; McFarland, E. W. *J. Am. Chem. Soc.* 2002, *124*, 12402.
[64] Kijima, T.; Yoshimura, T.; Uota, M.; Ikeda, T.; Fujikawa, D.; Mouri, S.; Uoyama, S. *Angewandte Chemie* 2004, *116*, 230.
[65] Feng, J.-J.; Xu, J.; Chen, H. *Electrochem. Commun.* 2006, *8*, 77.
[66] Feng, J.-J.; Hildebrandt, P.; Murgida, D. *Langmuir.* 2008, *24*, 1583.

[67] Feng, J.-J.; Li, A.-Q.; Wang, A.-J.; Lei, Z.; Chen, J.-R. *Microchim. Acta* 2011, *173*, 383.
[68] Vasa, P.; Lienau, C. *Angew. Chem. Int. Edit.* 2010, *49*, 2476.
[69] Wang, D.; Li, Y. *J. Am. Chem. Soc.* 2010, *132*, 6280.
[70] Tao, A.; Sinsermsuksakul, P.; Yang, P. *Nat Nano* 2007, *2*, 435.
[71] Tsuji, M.; Ogino, M.; Matsunaga, M.; Miyamae, N.; Matsuo, R.; Nishio, M.; Alam, M. J. *Cryst. Growth Des.* 2010, *10*, 4085.
[72] Wichner, N. M.; Beckers, J.; Rothenberg, G.; Koller, H. *J. Mater. Chem.* 2010, *20*, 3840.
[73] Song, Y.; Ma, Y.; Wang, Y.; Di, J.; Tu, Y. *Electrochim. Acta* 2010, *55*, 4909.
[74] Guo, S.; Zhai, J.; Fang, Y.; Dong, S.; Wang, E. *Chem. Asian J.* 2008, *3*, 1156.
[75] Sivakov, V. A.; Höflich, K.; Becker, M.; Berger, A.; Stelzner, T.; Elers, K.-E.; Pore, V.; Ritala, M.; Christiansen, S. H. *Chemphyschem* 2010, *11*, 1995.
[76] Srnová-Šloufová, I.; Vlčková, B.; Bastl, Z.; Hasslett, T. L. *Langmuir.* 2004, *20*, 3407.
[77] Gunawidjaja, R.; Kharlampieva, E.; Choi, I.; Tsukruk, V. V. *Small* 2009, *5*, 2460.
[78] Chen, X.; Cui, C.-H.; Guo, Z.; Liu, J.-H.; Huang, X.-J.; Yu, S.-H. *Small* 2011, *7*, 858.
[79] Lin, Y.-W.; Lin, Z.-H.; Huang, C.-C.; Chang, H.-T. *Multielemental Nanorods (Nanowires): Synthesis, Characterization and Analytical Applications*; ed.; Wiley-VCH Verlag GmbH and Co. KGaA, 2007.
[80] Tian, Z.-Q.; Ren, B.; Li, J.-F.; Yang, Z.-L. *Chem. Commun.* 2007, 3514.
[81] Chen, S.-Y.; Lazarides, A. A. *J. Phys. Chem. C* 2009, *113*, 12167.
[82] Keating, C. D.; Kovaleski, K. K.; Natan, M. J. *J. Phys. Chem. B* 1998, *102*, 9414.
[83] Li, J. F.; Huang, Y. F.; Ding, Y.; Yang, Z. L.; Li, S. B.; Zhou, X. S.; Fan, F. R.; Zhang, W.; Zhou, Z. Y.; WuDe, Y.; Ren, B.; Wang, Z. L.; Tian, Z. Q. *Nature* 2010, *464*, 392.
[84] Liu, F.; Cao, Z.; Tang, C.; Chen, L.; Wang, Z. *ACS Nano* 2010, *4*, 2643.
[85] Liu, F.; Tang, C.; Pan, J.; Cao, Z.; Wang, Z. *J. Phys. Chem. C* 2010, *114*, 9871.
[86] Yang, L.; Yan, B.; Premasiri, W. R.; Ziegler, L. D.; Negro, L. D.; Reinhard, B. M. *Adv. Funct. Mater.* 2010, *20*, 2619.
[87] Kubo, S.; Gu, Z.-Z.; Tryk, D. A.; Ohko, Y.; Sato, O.; Fujishima, A. *Langmuir* 2002, *18*, 5043.
[88] Lee, S. J.; Guan, Z.; Xu, H.; Moskovits, M. *J. Phys. Chem. C* 2007, *111*, 17985.
[89] Sezer, M.; Feng, J. J.; Khoa Ly, H.; Shen, Y.; Nakanishi, T.; Kuhlmann, U.; Hildebrandt, P.; Möhwald, H.; Weidinger, I. M. *Phys. Chem. Chem. Phys.* 2010, *12*, 9822.
[90] Chen, D.; Li, J. *Surf. Sci. Rep.* 2006, *61*, 445.
[91] Haes, A. J.; Zou, S.; Schatz, G. C.; Van Duyne, R. P. *J. Phys. Chem. B* 2003, *108*, 109.
[92] Wisitruangsakul, N.; Zebger, I.; Ly, K. H.; Murgida, D. H.; Ekgasit, S.; Hildebrandt, P. *Phys. Chem. Chem. Phys.* 2008, *10*, 5276.
[93] Zeravik, J.; Hlavacek, A.; Lacina, K.; Skládal, P. *Electroanalysis* 2009, *21*, 2509.
[94] Willner, I.; Shipway Andrew, N.; Willner, B. In *Molecules as Components of Electronic Devices*; Am. Chem. Soc., 2003; Vol. 844; pp 88.
[95] Tripp, R. A.; Dluhy, R. A.; Zhao, Y. *Nano Today* 2008, *3*, 31.
[96] Riskin, M.; Ben-Amram, Y.; Tel-Vered, R.; Chegel, V.; Almog, J.; Willner, I. *Anal. Chem.* 2011, *83*, 3082.
[97] Frasca, S.; von Graberg, T.; Feng, J.-J.; Thomas, A.; Smarsly, B. M.; Weidinger, I. M.; Scheller, F. W.; Hildebrandt, P.; Wollenberger, U. *ChemCatChem* 2010, *2*, 839.

[98] Dhand, C.; Das, M.; Datta, M.; Malhotra, B. D. *Biosens. Bioelectron.* 2011, *26*, 2811.
[99] Sun, T.; Qing, G. *Adv. Mater.* 2011, *23*, H57.
[100] Depagne, C.; Roux, C.; Coradin, T. *Anal. Bioanal. Chem.* 2010, *400*, 965.
[101] Stobiecka, M.; Hepel, M. *Sens. Actuators B: Chem.* 2010, *149*, 373.
[102] Stobiecka, M.; Coopersmith, K.; Hepel, M. *J. Colloid Interf. Sci.* 2010, *350*, 168.
[103] Kuo, C.-T.; Yu, J.-Y.; Huang, M.-J.; Chen, C.-H. *Langmuir* 2010, *26*, 6149.
[104] Sivanesan, A.; Ly, H. K.; Kozuch, J.; Sezer, M.; Kuhlmann, U.; Fischer, A.; Weidinger, I. M. *Chem. Commun.* 2011, *47*, 3553.
[105] Georg, S.; Kabuss, J.; Weidinger, I. M.; Murgida, D. H.; Hildebrandt, P.; Knorr, A.; Richter, M. *Phys. Rev. E* 2010, *81*, 046101.
[106] Reedy, C. J.; Elvekrog, M. M.; Gibney, B. R. *Nucleic. Acids. Res.* 2008, *36*, D307.
[107] Kranich, A.; Ly, H. K.; Hildebrandt, P.; Murgida, D. H. *J. Am. Chem. Soc.* 2008, *130*, 9844.
[108] Zuo, P.; Albrecht, T.; Barker, P. D.; Murgida, D. H.; Hildebrandt, P. *Phys. Chem. Chem. Phys.* 2009, *11*, 7430.
[109] Kranich, A.; Naumann, H.; Molina-Heredia, F. P.; Moore, H. J.; Lee, T. R.; Lecomte, S.; de la Rosa, M. A.; Hildebrandt, P.; Murgida, D. H. *Phys. Chem. Chem. Phys.* 2009, *11*, 7390.
[110] Ly•, H. K.; Marti•, M. A.; Martin, D. F.; Alvarez-Paggi, D.; Meister, W.; Kranich, A.; Weidinger, I. M.; Hildebrandt, P.; Murgida, D. H. *Chemphyschem* 2010, *11*, 1225.
[111] Feng, J.-J.; Murgida, D.; Kuhlmann, U.; Utesch, T.; Mroginski, M.; Hildebrandt, P.; Weidinger, I. *J. Phys. Chem. B* 2008, *112*, 15202.
[112] Wang, L.; Wang, E. *Electrochem. Commun.* 2004, *6*, 49.
[113] Hinnen, C.; Parsons, R.; Niki, K. *J. Electroanal. Chem. Interf. Electrochem.* 1983, *147*, 329.
[114] Reed, D. E.; Hawkridge, F. M. *Anal. Chem.* 1987, *59*, 2334.
[115] Wisitruangsakul, N.; Lenz, O.; Ludwig, M.; Friedrich, B.; Lendzian, F.; Hildebrandt, P.; Zebger, I. *Angew. Chem. Int. Edit.* 2009, *48*, 611.
[116] Willner, I.; Yan, Y. M.; Willner, B.; Tel-Vered, R. *Fuel Cells* 2009, *9*, 7.
[117] Millo, D.; Hildebrandt, P.; Pandelia, M.-E.; Lubitz, W.; Zebger, I. *Angew. Chem. Int. Edit.* 2011, *50*, 2632.
[118] Jiang, L.; Qian, J.; Cai, F.; He, S. *Anal. Bioanal. Chem.* 2011, *400*, 2793.
[119] Jiang, X.; Ataka, K.; Heberle, J. *J. Phys. Chem. C* 2007, *112*, 813.
[120] Park, B.-W.; Kim, D.-S.; Yoon, D.-Y. *Korean J. Chem. Eng.* 2011, *28*, 64.
[121] Feng, J.-J.; Murgida, D. H.; Kuhlmann, U.; Utesch, T.; Mroginski, M. A.; Hildebrandt, P.; Weidinger, I. M. *J. Phys. Chem. B* 2008, *112*, 15202.
[122] Thompson, W. R.; Pemberton, J. E. *Chem. Mater.* 1995, *7*, 130.
[123] Hench, L. L.; West, J. K. *Chem. Rev.* 1990, *90*, 33.
[124] Rossi, L. M.; Shi, L.; Quina, F. H.; Rosenzweig, Z. *Langmuir* 2005, *21*, 4277.
[125] Basiruddin, S. K.; Saha, A.; Pradhan, N.; Jana, N. R. *J. Phys. Chem. C* 2010, *114*, 11009.
[126] [126] Selvan, S. T.; Tan, T. T. Y.; Yi, D. K.; Jana, N. R. *Langmuir* 2009, *26*, 11631.
[127] Yang, W.; Lopez, P. J.; Rosengarten, G. *Analyst* 2011, *136*, 42.
[128] Farrell, R. A.; Petkov, N.; Morris, M. A.; Holmes, J. D. *J. Colloid Interf. Sci.* 2010, *349*, 449.

[129] Bantz, K. C.; Meyer, A. F.; Wittenberg, N. J.; Im, H.; Kurtulus, O.; Lee, S. H.; Lindquist, N. C.; Oh, S.-H.; Haynes, C. L. *Phys. Chem. Chem. Phys.* 2011, *13*, 11551.
[130] Millo, D.; Pandelia, M.-E.; Utesch, T.; Wisitruangsakul, N.; Mroginski, M. A.; Lubitz, W.; Hildebrandt, P.; Zebger, I. *J. Phys. Chem. B* 2009, *113*, 15344.
[131] Lee, K.; Drachev, V. P.; Irudayaraj, J. *ACS Nano* 2011, *5*, 2109.
[132] Huang, J.-Y.; Zong, C.; Xu, L.-J.; Cui, Y.; Ren, B. *Chem. Commun.* 2011, *47*, 5738.
[133] Matthäus, C.; Bird, B.; Miljkovic, M.; Chernenko, T.; Romeo, M.; Diem, M. In *Method. Cell Biol.*; Dr. John, J. C.; Dr. H. William Detrich, III, Eds.; Academic Press, 2008; Volume 89; pp 275.
[134] Millo, D.; Harnisch, F.; Patil, S. A.; Ly, H. K.; Schröder, U.; Hildebrandt, P. *Angew. Chem. Int. Edit.* 2011, *50*, 2625.
[135] Baldwin, K. J.; Batchelder, D. N. *Appl. Spectrosc.* 2001, *55*, 517.
[136] Genov, D. A.; Sarychev, A. K.; Shalaev, V. M.; Wei, A. *Nano Lett.* 2003, *4*, 153.
[137] He, L.; Smith, E. A.; Natan, M. J.; Keating, C. D. *J. Phys. Chem. B* 2004, *108*, 10973.
[138] Zhang, L. *Biosens. Bioelectron.* 2008, *23*, 1610.
[139] Song, Y.; Wan, L.; Cui, K.; Liu, L.; Zhang, M.; Liao, J.; Wang, L.; Li, Z. *J. Electroanal. Chem.* 2011, *656*, 206.
[140] Ramírez, P.; Andreu, R.; Cuesta, Á.; Calzado, C. J.; Calvente, J. J. *Anal. Chem.* 2007, *79*, 6473.
[141] Juskowiak, B. *Anal. Bioanal. Chem.* 2011, *399*, 3157.
[142] Yang, Y.; Zhao, L. *TrAC, Trends Anal. Chem.* 2010, *29*, 980.
[143] Deiss, F.; Sojic, N.; White, D.; Stoddart, P. *Anal. Bioanal. Chem.* 2010, *396*, 53.
[144] Wabuyele, M.; Yan, F.; Vo-Dinh, T. *Anal. Bioanal. Chem.* 2010, *398*, 729.
[145] He, Y.; Su, S.; Xu, T.; Zhong, Y.; Zapien, J. A.; Li, J.; Fan, C.; Lee, S.-T. *Nano Today* 2011, *6*, 122.
[146] Lo, H.-C.; Hsiung, H.-I.; Chattopadhyay, S.; Han, H.-C.; Chen, C.-F.; Leu, J. P.; Chen, K.-H.; Chen, L.-C. *Biosens. Bioelectron.* 2011, *26*, 2413.
[147] Graham, D.; Stevenson, R.; Thompson, D. G.; Barrett, L.; Dalton, C.; Faulds, K. *Faraday. Discuss.* 2011, *149*, 291.
[148] Sun, S.; Thompson, D.; Schmidt, U.; Graham, D.; Leggett, G. J. *Chem. Commun.* 2010, *46*, 5292.
[149] Dougan, J. A.; MacRae, D.; Graham, D.; Faulds, K. *Chem. Commun.* 2011, *47*, 4649.
[150] Maiti, K. K.; Samanta, A.; Vendrell, M.; Soh, K.-S.; Olivo, M.; Chang, Y.-T. *Chem. Commun.* 2011, *47*, 3514.
[151] Hong, G.; Tabakman, S. M.; Welsher, K.; Chen, Z.; Robinson, J. T.; Wang, H.; Zhang, B.; Dai, H. *Angew. Chem. Int. Edit.* 2011, *50*, 4644.
[152] Sanda, B.; Dumitrita, R.; Adela P.; Lucian B.-T.; Simion A. *Nanotechnology* 2011, *22*, 055702.
[153] Serra, A.; Manno, D.; Filippo, E.; Buccolieri, A.; Urso, E.; Rizzello, A.; Maffia, M. *Sens. Actuators B: Chem.* 2011, *156*, 479.
[154] El-Said, W. A.; Kim, T.-H.; Kim, H.; Choi, J.-W. *Biosens. Bioelectron.* 2010, *26*, 1486.
[155] Niu, J. J.; Schrlau, M. G.; Friedman, G.; Gogotsi, Y. *Small* 2011, *7*, 540.
[156] Lin, S.-Y.; Chen, N.-T.; Sun, S.-P.; Chang, J. C.; Wang, Y.-C.; Yang, C.-S.; Lo, L.-W. *J. Am. Chem. Soc.* 2010, *132*, 8309.

In: Plasmons: Structure, Properties and Applications
Editors: Anton E. Turunen and Johannes O. Niemi
ISBN: 978-1-62100-060-0

Chapter 3

DISPERSION AND DAMPING OF PLASMONIC EXCITATIONS IN METAL/METAL INTERFACES

Antonio Politano*
Università degli Studi della Calabria, Dipartimento di Fisica,
Rende (Cs), Italy

ABSTRACT

We present high-resolution electron energy loss spectroscopy (HREELS) measurements on surface plasmon dispersion in systems exhibiting quantum well states, i.e. Na/Cu(111), Ag/Cu(111), and Ag/Ni(111). Our results demonstrate that the dominant coefficient of surface plasmon dispersion for thin and layer-by-layer Ag films presenting quantum well states is quadratic even at small $q_{\|}$, in contrast with previous measurements on Ag semi-infinite media and Ag thin films deposited on Si(111). We suggest that this behavior is due to screening effects enhanced by the presence of quantum well states shifting the position of the centroid of the induced charge less inside the geometrical surface compared to Ag surfaces and Ag/Si(111). For ultrathin Ag films, i.e. two layers, the dispersion was found to be not positive, as theoretically predicted. Annealing caused an enhancement of the free-electron character of the quantum well states, thus inducing a negative linear term of the dispersion curve of the surface plasmon. Moreover, we report the first experimental evidence of chemical interface damping in thin films for K/Ag/Ni(111). As regards Na/Cu(111), we found a different dispersion curve compared to thick Na films, thus confirming the enhanced screening by Na quantum well states.

Results reported here should shed light on the influence of quantum well states on dynamical screening phenomena in thin films.

INTRODUCTION

In low-dimensional systems, the reduced dimensionality may induce the appearance of novel electronic states. In semi-infinite media (surfaces), surface states are formed within

* E-mail:antonio.politano@fis.unical.it

energy gaps of the projected bulk band structure due to the broken symmetry at the solid/vacuum interface. Electrons occupying such states are confined at the top layer of the bulk sample and form a two-dimensional nearly free-electron gas. On the other hand, in thin films discrete states are formed, characterized by a notable dependence of their energy on thickness. Such states, called quantum well states (QWS), describe standing electron waves confined within the film.

Recent experimental [1-7] studies have demonstrated the existence of variations with film thickness for properties such as the electronic density of states, electron-phonon coupling, chemical reactivity, superconductivity, magnetism, surface energy, and thermal stability.

Further information on the physical and chemical properties of thin films can be obtained from their electronic collective excitations and, in particular, from the dispersion relation of the surface plasmon (SP). Nevertheless, the influence of electron quantum confinement on the dispersion relation of electronic collective excitations in nanoscale thin films has not been investigated yet. Moreover, the different distribution of occupied and unoccupied electronic states in thin films with respect to surfaces would imply enhanced damping mechanisms for collective excitations by creating electron-hole pairs.

Herein high-resolution electron energy loss spectroscopy (HREELS) measurements of systems exhibiting electron quantum confinement, i.e. Ag/Ni(111), Ag/Cu(111), and Na/Cu(111) are reported. Our results should shed light on the influence of QWS on dynamical screening phenomena in thin films.

FORMATION OF QWS

The electronic structure of films has been widely investigated by photoemission spectroscopy in recent years. These studies focused particularly on two main spectral features: the occupied Shockley-type surface state resulting from the termination of the crystal [8], and the QWS due to the quantum confinement of the *sp* valence electrons in the adlayer [9-13]. The binding energy of the Shockley state depends critically on film thickness.

Figure 1 summarizes experimental photoemission data acquired at 300 K for Ag/Cu(111) by Mathias et al. [14] as a function of silver film thickness. Moreover, the film morphology was found to be a very important parameter influencing both the electronic structure. Recently, it has been demonstrated that the dispersion of QWS changes upon annealing the adlayer [15]. Angle-resolved photoemission experiments showed that Ag QWS on Au(111) have flat in-plane dispersion in a disordered film and a nearly free-electron-like dispersion in an annealed and well-ordered film. Accordingly, the sp density of states of the film may be tuned by annealing. As an example, Figure 2 shows photoemission data recorded at room temperature for a silver film (of nominally 15 ML) on Cu(111), both before and after the annealing procedure. The peak position of QWS shifted slightly to a higher binding energy. Moreover, annealing reduced the broadness of the peak as a consequence of a higher flatness of the adlayer (Figure 3).

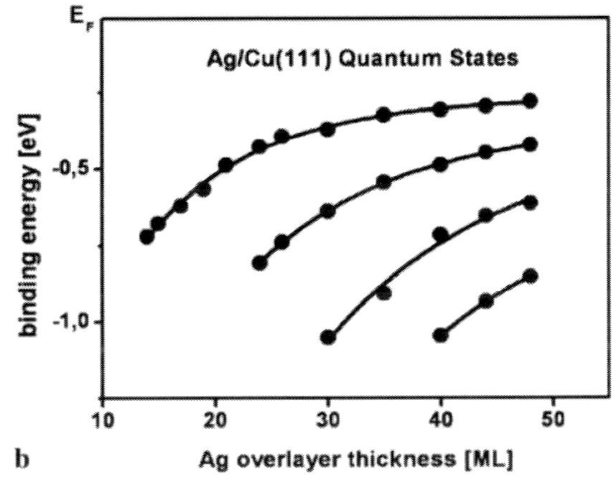

Figure 1. Energy of the $v = 1$ to $v = 4$ QWS and resonances as a function of nominal film thickness as determined from photoemission spectra from non-annealed silver films (taken from Ref. 14, with permission).

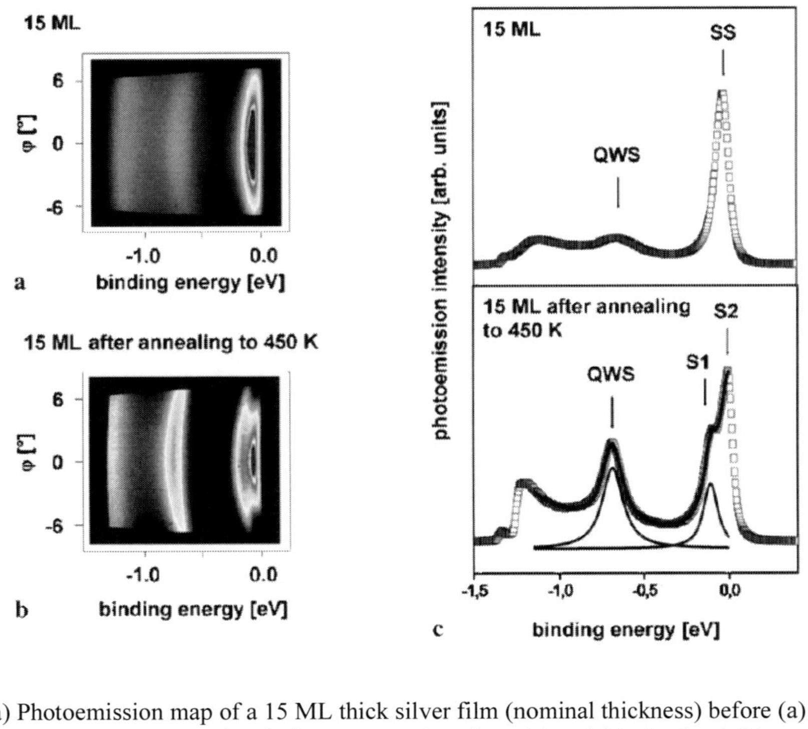

Figure 2. (a) Photoemission map of a 15 ML thick silver film (nominal thickness) before (a) and after (b) annealing at 450 K; (c) normal emission spectra taken from (a) and (b) clearly visible are the modifications in the spectral shape of the quantum wells state (QWS) and in the energy region, where the Shockley surface state (SS) is observed. *S*1 and *S*2 indicate the two distinct Shockley surface states appearing in the spectrum after annealing; the two Lorentzians in the *bottom graph* result from a fit into the QWS-peak and *S*1-peak of the bottom spectrum. Also shown as a *solid line* is the total fit result. A Gaussian was used to reproduce the low energy tail of the spectrum and a Lorentzian to fit the peak *S*2 (taken from Ref. 14 with permission).

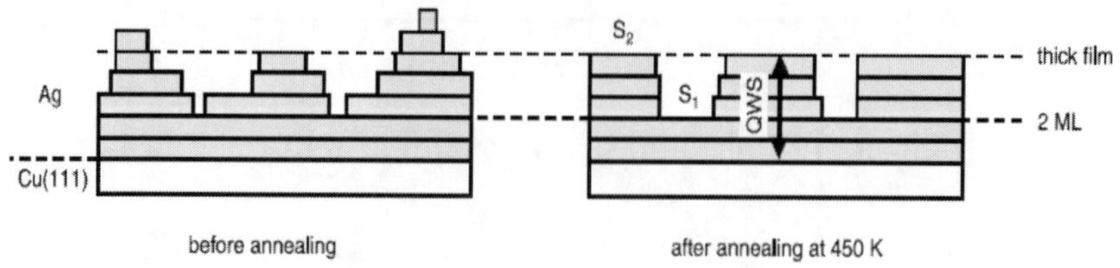

Figure 3. Schematic summary of the morphological changes induced by the heat treatment of the silver film (taken from Ref. 14 with permission).

COLLECTIVE EXCITATIONS

Theoretical Background

In the case of a metallic slab there are two surfaces with coupled charge density oscillations. This coupling produces new class of surface charge oscillations with well defined parity, odd and even, so the dispersion relation of SPs has two branches: above and below ω_S, respectively. The microscopic description of such surface excitations in metallic slabs, which includes both collective and single-particle excitations, can be obtained using surface spectral functions

$$D_p(Q,\omega) = \frac{c_p - c_{-p}\epsilon_p(Q,\omega)}{1+\epsilon_p(Q,\omega)}, \quad c_p = 1 - pe^{-QL}, \quad p = \pm$$

where $\varepsilon_p(Q,\omega)$ is the surface dielectric function which can be calculated in Random Phase Approximation [16-18]. Spectral intensity of odd and even surface excitations is shown in Figure 4. Peaks correspond to two surface plasmons.

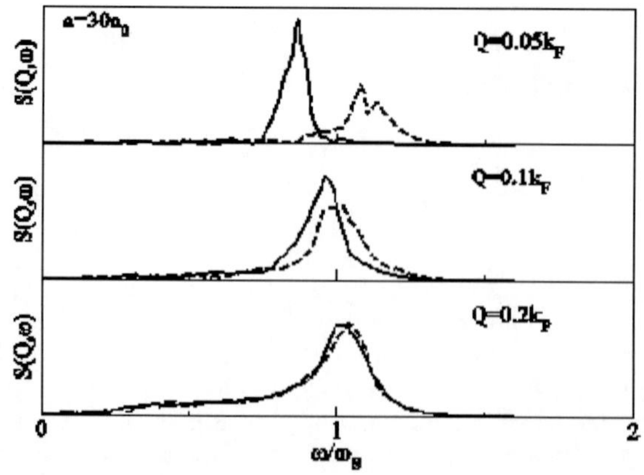

Figure 4. Spectral of even (full line) and odd (dashed line) surface excitations for metallic slab of thickness a = $30a_0$ and density is r_s=2.0. Parallel wave vectors are a) Q = 0.05kF; b) Q = $0.1k_F$; c) Q = $0.2k_F$ (taken from Ref 17 with permission).

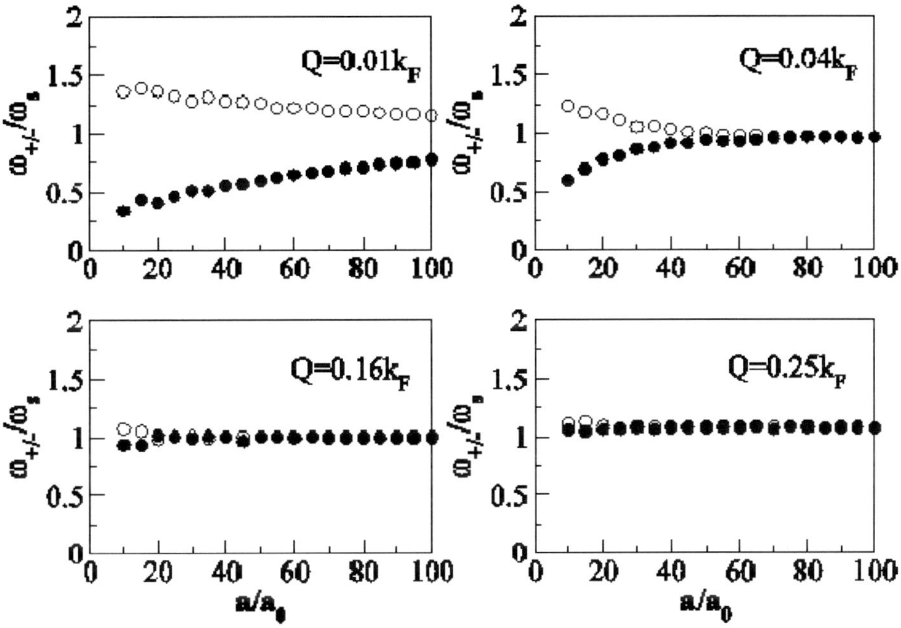

Figure 5. Collective mode frequencies against the slab thickness for various wave-vectors. Dots are for even modes, and circles for odd modes (taken from Ref 17 with permission).

For thicker slabs or larger wave vectors, charge density oscillations become uncoupled, and plasmon frequencies degenerate. For example, Figure 5 shows that SPs for Q = 0.25 k_F are degenerated for slab thicknesses from 10 to 100 a_0. Long-wavelength surface excitations in thin metallic slabs exhibit some interesting features with completely quantum mechanical character due to the presence of QWS [19] and also the electron-hole (Q, ω) phase space becomes quantized.

The phase space (Q, ω) for pair excitation from QW state with quantum number n to QW states with quantum number m is restricted by curves (Figure 6).

$$\omega_{mn}^{min} = \Delta_{mn} + \frac{Q^2}{2m} - v_n Q$$

$$\omega_{mn}^{max} = \Delta_{mn} + \frac{Q^2}{2m} + v_n Q$$

where $\Delta_{mn} = E_m - E_n,\ E_n < E_F$ and $v_n = \frac{\sqrt{2m(E_F - E_n)}}{m}$ is maximum phase velocity of electron with quantum number n.

QWS obviously induce gaps in (Q, ω) phase space, and their widths decrease with wave vector and frequency (dotted vertical lines in Figure 6). Accordingly, if a long-wavelength plasmon occurs in a gap it cannot decay by exciting an electron from lower to higher QWS. But for thicker slabs (a >20 a_0) and for $\omega \approx \omega_S$ gaps disappear almost in all wave-vector ranges and SPs are damped even for Q≈ 0. For example, SP is significantly damped for Q=0.05 k_F (Figure 7a) but also for Q=0.02 k_F (Figure 7c). High precision of the calculations enables the analysis not only of the collective (SP) modes but also of the contributions to spectra due to the transition between QWS (interband transitions). The quantized phase space for the fixed wave vector (full vertical lines in Figure 6) predicts that width of these structures increases with the wave vector. Figures 7b) and d) show sharp periodic peaks corresponding to transition between successively unoccupied n = N, N-1, N-2,.. and first occupied n = N + 1 QWS. The intensity of the field which is produced by such transitions is strongly screened by SPs and this is the reason why they are not visible in Figures 7a) and c).

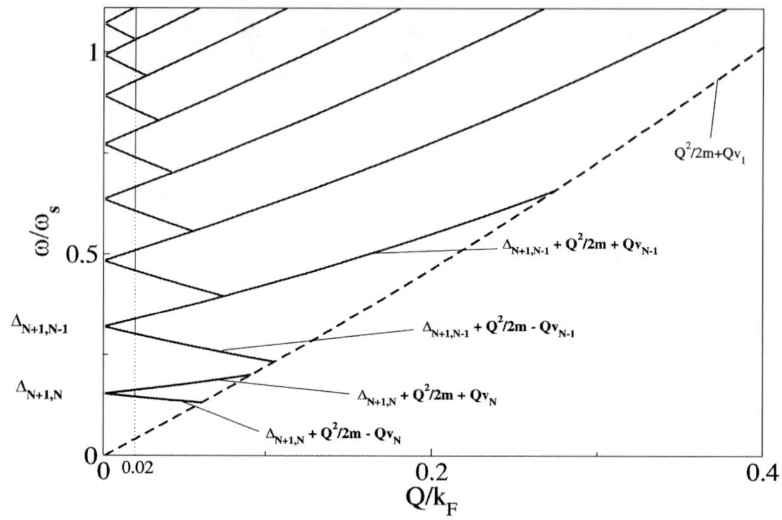

Figure 6. Phase space of interband n → N + 1; n = 1, 2,…N (between full lines) and intraband n = 1 → n = 1 (below dashed line) transitions for metallic slabs of thicknesses a = 30 a_0 and r_s = 2.0. N is number of occupied states in perpendicular direction (taken from Ref 19 with permission).

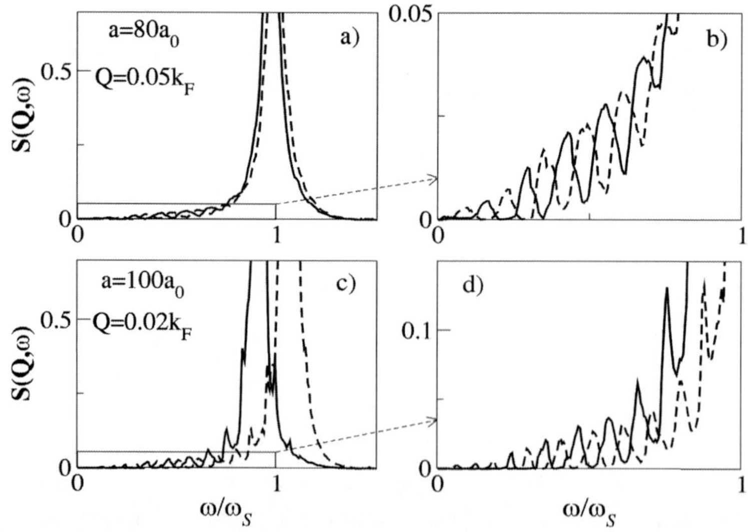

Figure 7. Details of surface excitation spectra showing single-particle excitations. Full line is for even excitations, and dashed line for odd excitations (taken from Ref. 19 with permission).

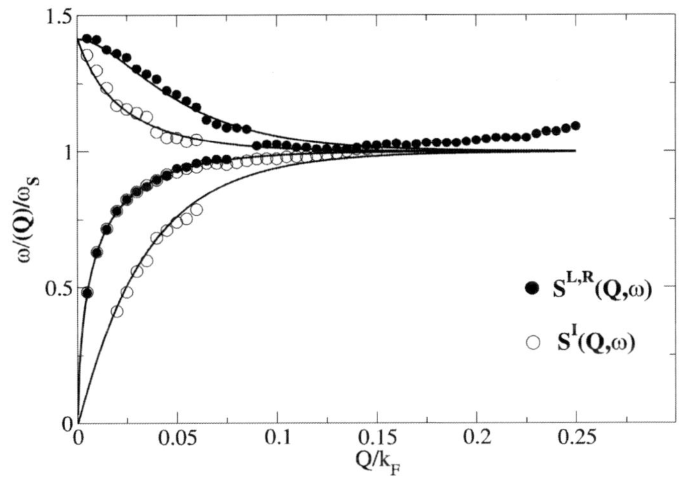

Figure 8. Dispersion relation of SPs for pair of coupled metallic slabs (a_1=a_2=30 a_0, r_{s1}=r_{s2}=2 and D = 25 a_0). Full line presents dispersion relations in long-wavelength limits, circles/dots presents surface spectral function maximums (taken from Ref. 20 with permission).

For a metallic slab near semi-infinite metallic substrate, charge density fluctuations on all three surfaces also become coupled and plasmon frequency is split in three new frequencies. Similarly, for two parallel metallic slabs there are four SP branches [20]. Figure 8 shows dispersion relations of SPs in two metallic slabs. Results are also compared with calculations in long-wavelength limit when frequency splitting is significant. In the short-wavelength limit when quantum mechanical effects appear (positive slope and Landau damping), SPs already become independent (uncoupled).

Experimental

Measurements were carried out in a UHV chamber operating at a base pressure of 5×10^{-9} Pa, equipped with standard facilities for surface characterizations. HREEL experiments were performed by using an electron energy loss spectrometer (Delta 0.5, SPECS). The samples were single-crystal surfaces of Ni(111) and Cu(111) with a purity of 99.9999%. Surfaces were cleaned by repeated cycles of ion sputtering and annealing at 800-900 K. Surface cleanliness and order were checked using Auger electron spectroscopy measurements and low-energy electron diffraction (LEED), respectively. Silver was deposited onto the Ni(111) and Cu(111) surface by evaporating from an Ag wire wrapped on a tungsten filament. Well-ordered Ag films could be obtained at very low deposition rates (≈0.05 ML/min). Alkalis (Na, K) were evaporated in the UHV chamber by means of a well outgassed dispenser (Saes Getters). The occurrence of the p(1x1)-Ag, (3/2x3/2)-Na, and p(2x2)-K LEED patterns was used as the calibration point of θ_{Ag}=1.0 ML θ_{Na}=0.44 ML, and θ_K=0.25 ML, respectively. A constant sticking coefficient was assumed to obtain other desired coverage. The energy resolution of the spectrometer was degraded to 7 meV so as to increase the signal-to-noise ratio for off-specular spectra. The angular acceptance α of our electron analyzer was ±0.5°. All depositions and measurements were made at room temperature.

Results and Discussion

Ag Films on Ni(111)

Ag surfaces are characterized by a strong lowering of the SP energy, which follows a positive dispersion as a function of the parallel momentum transfer. Such behavior was ascribed to the presence of filled d bands [21-23]. However, experimental studies on low-dimensional Ag systems, such as ultrathin films on metal [24] and semiconductor [25-27] substrates, nanowires [28] or nanoparticles [29,30] are less common.

Ag on Ni(111) offers the possibility of investigating the relationship between quantum electron confinement and the damping of collective excitations in ultrathin films as confined electron states in the Ag adlayer exist. As a matter of fact, photoemission measurements [31] revealed the occurrence of Ag QWS for films with thickness from 0 to 15 layers. Ag on Ni(111) is an example of film growth in large (16%) mismatched materials. Such a large misfit determines the silver film to have the crystalline structure of bulk silver even for ultrathin Ag layers [32]. Moreover, the small solubility of Ag into Ni prevents from the dissolution of the silver layer into the nickel substrate [33, 34]. As a matter of fact, the formation of a surface alloy was not reported for this system [5, 36]. For coverages higher than three layers, a reconstruction of the silver overlayer was observed [32]. The deposited Ag layer exhibits a 7x7 moiré structure [33].

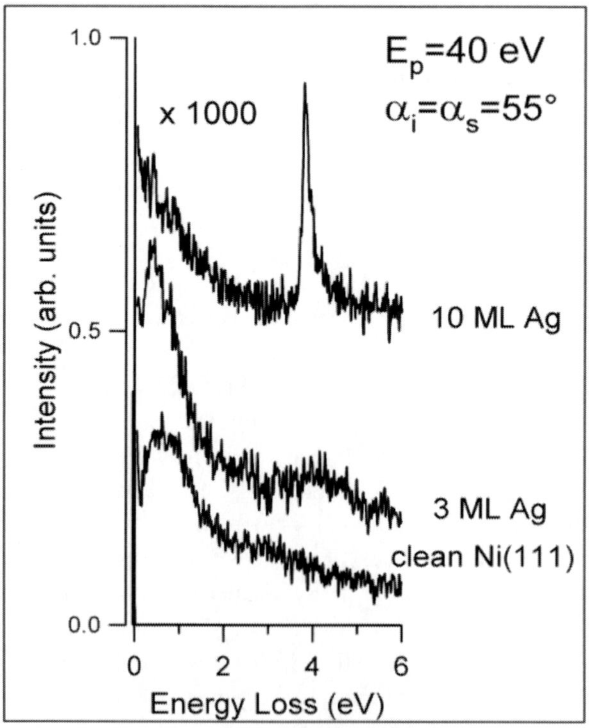

Figure 9. HREEL spectra for Ag/Ni(111) as a function of Ag coverage.

Figure 9 shows HREEL spectra for thin Ag layers on Ni(111) as a function of Ag coverage. The spectrum of the Ni(111) surface is characterized by a broad peak at 1.0 eV. For less than 2 ML of Ag, the spectrum is extremely broad without a well-defined peak. A broad Ag SP at 4.2 eV arose for Ag coverages above 2 ML. However, its plasma energy was considerably higher than that of the SP of semi-infinite Ag, as also found for Ag/Si(111) [25, 26]. As the Ag coverage is increased, the energy of the plasmonic mode reduced and its line shape became sharper.

Such finding has to be ascribed to the s-d polarization. For thin Ag films, the overall screening of the plasmon via the polarizable d electron medium diminishes and a higher SP frequency occurs [21]. Moreover, the spill-out region not affected by s-d polarization becomes more important, causing a further blue-shift of the SP frequency.

Important information on the screening properties of metal systems is given by the dispersion relation of the SP.

To measure plasmon dispersion, values for the parameters E_p, impinging energy, and θ_i, the incident angle, were chosen so as to obtain the highest signal-to-noise ratio. The primary beam energy used for the dispersion, E_p=40 eV, provided, in fact, the best compromise among surface sensitivity, the highest cross-section for the plasmonic excitation, and $q_{||}$ resolution. As,

$$\hbar \vec{q_{||}} = \hbar \left(\vec{k_i} \sin\theta_i - \vec{k_s} \sin\theta_s \right)$$

the parallel momentum transfer $q_{||}$ is determined by the values of E_p, E_{loss}, θ_i and θ_s:

$$q_{||} = \frac{\sqrt{2mE_p}}{\hbar}\left(\sin\theta_i - \sqrt{1 - \frac{E_{loss}}{E_p}}\sin\theta_s\right)$$

where E_{loss} is the energy loss and θ_s is the electron scattering angle [22].

Hence, it is possible to estimate the integration window in reciprocal space [22]:

$$\Delta q_{||} \approx \frac{\sqrt{2mE_p}}{\hbar}\left(\cos\theta_i + \sqrt{1 - \frac{E_{loss}}{E_p}}\cos\theta_s\right)\cdot\alpha$$

where α is the angular acceptance of the apparatus [2]. Under our experimental conditions, $\Delta q_{||}$=0.012 Å^{-1}, much less than the scanned range in the reciprocal space.

Selected HREEL spectra for 10 ML of Ag as a function of the scattering angle are showed in figure 10. The Ag SP energy dispersed from 3.751 up to 3.880 eV.

The measured dispersion curve $E_{loss}(q_{||})$ in Figure 11 was fitted by a second-order polynomial given by:

$$E_{loss}(q_{||})=A+Bq_{||}+Cq_{||}^2$$

where A=(3.751±0.002) eV, B=(0.000±0.004) eV·Å, and C=(1.57±0.04) eV·Å^2.

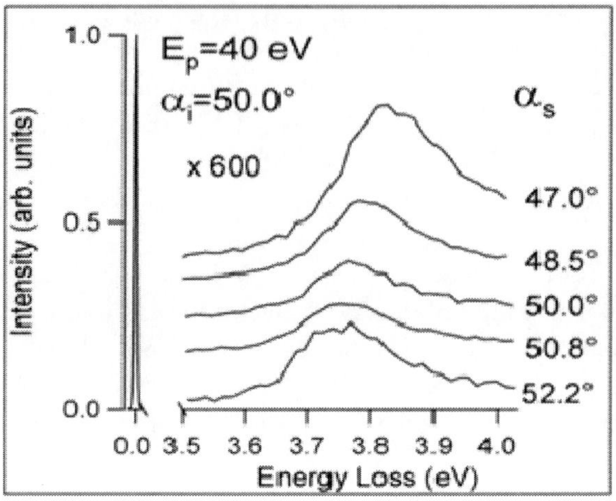

Figure 10. HREEL spectra as a function of the scattering angle for 10 ML Ag/Ni(111) at T=300 K.

Hence, the dispersion curve for 10 ML Ag/Ni(111) reported in Figure 10 is purely quadratic, as the linear coefficient B is null. Similar results were obtained for slightly different Ag thicknesses (Figure 10) for which QWS were also observed to exist by photoemission spectroscopy [31]. At higher thicknesses at which the discrete electronic structure of QWS [31] is vanishing, the dispersion curve deviates from being purely quadratic. As expected [21, 25, 26], the energy of the SP changed with Ag coverage.

It is worth mentioning that great efforts have been devoted to solve the controversy concerning the quadratic versus linear form of the dispersion curve. The conclusion of this long debate [22, 37-40] is that the dominant coefficient of the SP dispersion for small momenta is always linear. On the other hand, quadratic terms become important at higher values of the parallel transfer momentum $q_{||}$. However, quadratic terms are nearly absent for Ag(100) [22, 38].

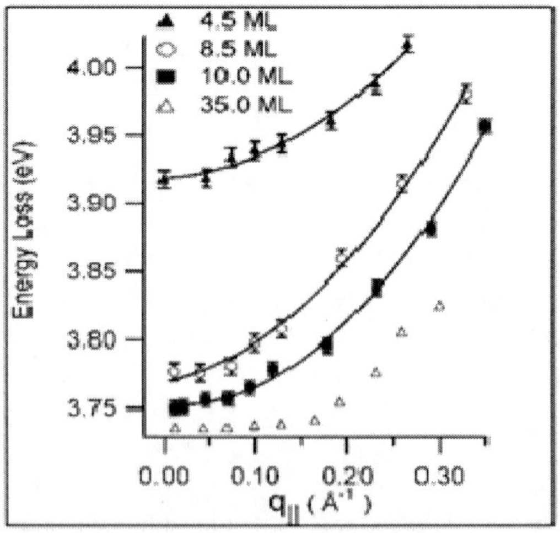

Figure 11. SP dispersion as a function of the parallel transfer momentum for 4.5, 8.5, 10.0, and 35.0 ML Ag/Ni(111) at T=300 K. The solid line indicates the best-fit curve for experimental data.

According to Feilbelman's model [41] of the SP dispersion, a direct relation between the linear coefficient of the dispersion curve and the position of the centroid of the induced charge, $d_\perp$, may be found [39]. In Ag such centroid lies inside the geometrical surface (z<0). The values of the dispersion coefficients and of the calculated position of $d_\perp$ for Ag single-crystal surfaces, 10 ML Ag/Ni(111), and 10 ML Ag/Si(111) are reported in Table 1. As the linear coefficient B increased, $d_\perp$ became more negative. On the contrary, no particular relationship involving the quadratic coefficient C exists.

Feibelman [42] considered the relation between the SP dispersion and the 4d-5s excitations induced by the field of the SP. He argued that the linear coefficient is increased in Ag(100) because the 4d electrons lie closer to the centroid of the oscillating free-electron charge. On the basis of this result, we suggest that for 10 ML Ag/Ni(111) the linear coefficient is null as the centroid of the induced charge of the SP lies much farther with respect to 4d electrons compared to other Ag systems. We ascribe this effect to screening processes [43] enhanced by the existence of Ag QWS [31].

Table 1. Best-fit values A, B, and C for different Ag surfaces [38], 10 ML Ag/Ni(111) (our data), and 10 ML Ag/Si(111) [26]. The values of $d_\perp$ were calculated by using Eq. 4 of Ref. 39. The values for Ag surfaces were reported in the same Ref. 39

	A	B	C	$d_\perp$ (Å)
10 ML Ag/Ni(111)	3.751	0.00	1.57	-0.6
Ag(110), <1 $\bar{1}$ 0>	3.683	0.14	3.60	-0.7
10 ML Ag/Si(111)	3.796	0.35	2.56	-0.9
Ag(111)	3.684	0.45	3.40	-1.3
Ag(110), <001>	3.681	0.65	3.10	-1.9
Ag(100)	3.690	1.40	0.06	-2.6

The presence of confined electronic states in the adlayer modifies the shape of the potential barrier that becomes smoother. Very likely, dynamic screening would be more efficient for Ag thin films deposited onto metal substrates with respect to semiconductor surfaces. As a matter of fact, for Ag/Si(111) the linear coefficient of the SP dispersion was decreasing with thickness but it was always different from zero [26]. Moreover, hybridization effects should also be taken into account, as recently found for Ag films on Ge(111) with thickness up to 9 ML [44]. Their influence should be different for different substrates. Reflection (phase shifts) of the QWS wave functions depends on the interfacial hybridization with the electronic states of the substrate, thus influencing the dynamic screening properties and the electronic response of the Ag/Ni(111) interface.

One advantage of EELS technique is that changing the primary beam energy of the electrons, it is possible to modify their mean free path in the solid and thus their penetration length [45]. This allows direct control of the surface sensitivity. The energy loss position of the Ag SP was found to shift as a function of the primary electron beam energy (Figure 12a).

This could be ascribed at the combined effect of the dispersion in the parallel transfer momentum and changes in the penetration length of the primary electrons.

Interestingly, both the Ag SP energy ω_{sp} and the full width at half maximum (FWHM) follow a linear relation as a function of the inverse of the primary beam energy E_p (Figure 12 b).

$$\omega_{sp}=a+b*(1/E_p)$$

$$FWHM= c+d*(1/E_p)$$

where a=3.73 eV; b= 4.39 eV^2; c=0.11 eV; and d= 3.18 eV^2.

Increasing the kinetic energy of incident electrons from 20 to 170 eV caused a 200 meV shift of the energy loss position. Furthermore, the FWHM decreased by 140 meV for increasing primary energy. The FWHM of the Ag SP is associated with the damping of the SP, as a results of its decaying into single-particle excitations (Landau damping) through 5sp-5sp or 4d-5sp band transitions.

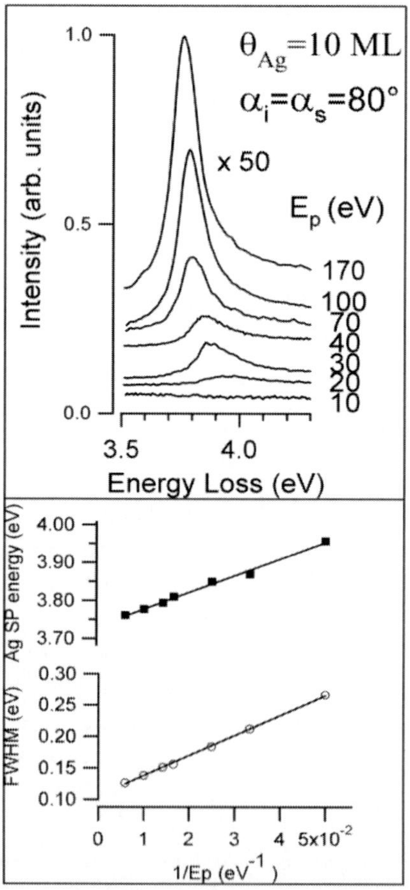

Figure 12. (a) HREEL spectra for 10 ML Ag/Ni(111) as a function of the primary electrons beam energy. (b) (■) Loss energy and (○) FWHM as a function of the inverse of the kinetic energy of primary electrons.

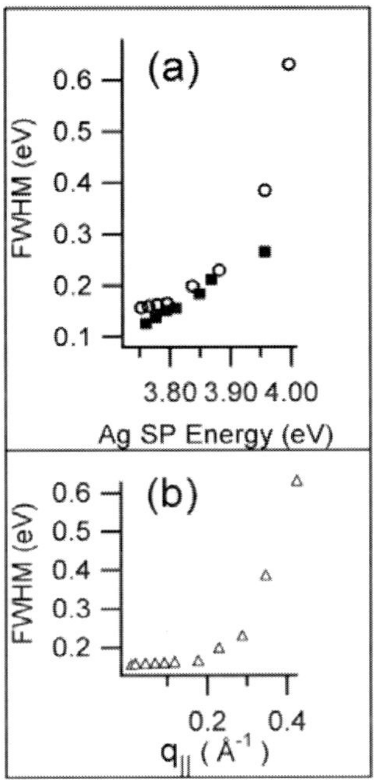

Figure 13. (a) FWHM of the Ag SP peak for 10 ML Ag/Ni(111) as a function of the energy of the Ag SP, as obtained in different experimental conditions (■) $\theta_i=\theta_s$ =55° fixed and E_p varying (○)θ_i =55°, E_p =40 eV fixed and θ_s varying. (b) FWHM of the SP peak for 10 ML Ag/Ni(111) as a function of the parallel momentum transfer $q_{||}$, calculated using with the following experimental parameters: θ_i =55°, E_p =40 eV. The abrupt increase of the FWHM beyond $q_{||}$=0.2 $Å^{-1}$ indicates that an extra decay channel due to indirect single-particle transition between surface electronic states opens up.

The FWHM of the SP peaks is plotted as a function of Ag SP energy in Figure 13a. The FWHM was found to continuously increase a function of Ag SP energy and no minimum exists (figure 13a). The loss width notably grows beyond 3.80 eV (empty circles, primary beam energy fixed) indicating the opening of a new efficient decay channel. However, the same plot obtained by changing the impinging beam energy (filled squares) did not present such critical energy. This result was ascribed to effects caused by differences in penetration length of the impinging electrons.

The behavior of the FWHM as a function of the parallel transfer momentum $q_{||}$ (figure 13b) demonstrated that for $q_{||}$>0.2 $Å^{-1}$ electrons may be promoted from occupied to unoccupied electronic states such that the corresponding plasmon peak would broaden considerably until decaying into the single-particle excitation continuum.

The width of the SP peak at $q_{||}$=0 is a sensitive function of the lattice potential, i.e. it is influenced by the periodic potential of the bulk. Such decay mechanisms can be direct or mediated by the exchange of reciprocal lattice vectors or phonons [22]. However, at larger values of $q_{||}$ the plasmon lifetime becomes less sensitive to the bulk lattice potential and the increasing of the damping is then caused by electron-hole pair excitations in the surface region. In real systems the lifetime of the plasmon is further limited by the scattering against crystallographic defects. For a thinner film, extra decay channels exist as compared with a thicker film. As a matter of fact, sp-sp interband transitions were found [26, 46] to be more efficient in thinner films rather than in thicker ones.

A comparison of the behavior of the FWHM of the Ag SP in 10 ML Ag/Ni(111), 10 ML Ag/Si(111), and Ag(111) is reported in Figure 14. Panel (a) shows the FWHM dependence on the Ag SP energy, while panel (b) on the parallel transfer momentum. For 10 ML Ag/Si(111), Yu et al. [26] found for different thicknesses an initial decrease of the FWHM as a function of the Ag SP energy, followed by an abruptly increasing. Such abrupt increase was ascribed to the opening of a new damping channel, i.e. intraband transitions between Ag 5sp-derived QWS. The initial decrease of FWHM was associated with the increased surface barrier due to the ionic pseudopotential of the crystal.

On the contrary, the FWHM of Ag SP in 10 ML Ag/Ni(111) has an initial flat dependence on both Ag SP energy and $q_{||}$ and a stronger dependence beyond a critical value of both Ag SP energy and $q_{||}$. The notable difference existing in the behavior of the FWHM of Ag SP in 10 ML Ag deposited on Si(111) and on Ni(111) may be tentatively ascribed to substrate effects. The FWHM of the Ag SP in Ag(111) [38] has, in analogy with 10 ML Ag/Ni(111), an initial flat dependence on both Ag SP energy and $q_{||}$. The critical values of Ag SP and $q_{||}$ were found to be 3.78 eV and 0.15 Å^{-1}, respectively. Beyond such values, the FWHM significantly increased. It is worth to stress that the FWHM of the Ag SP in Ag(111) was always notably lower than that of Ag SP in thin films.

This finding is ascribed to enhanced damping processes via 5sp-5sp indirect transitions, due to the different electronic properties of thin films with respect to surfaces.

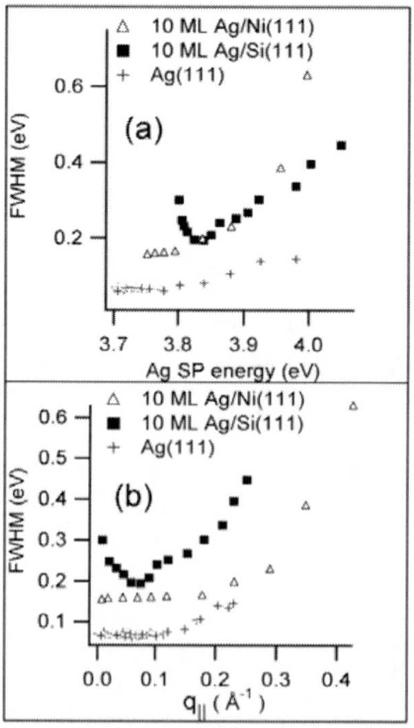

Figure 14. (a) FWHM of the Ag SP peak as a function of the energy of the Ag SP for (Δ) 10 ML Ag/Ni(111); (■) 10 ML Ag/Si(111) (data taken from ref. 26); and (+) Ag(111) (data taken from ref. 38) (b) FWHM of the SP peak as a function of the parallel momentum transfer $q_{||}$ for (Δ) 10 ML Ag/Ni(111); (■) 10 ML Ag/Si(111) (data taken from ref. 26); and (+) Ag(111) (data taken from ref. 38).

If the surface is exposed to chemically reactive atoms or molecules, the distribution of occupied and unoccupied electronic states changes. Accordingly, differences in damping processes and in the energy of the plasmonic excitation are quite expected. Rather than considering the new overlayer plasmon, we investigated the overlayer-induced modification of the substrate plasmon. Figure 15a shows Ag surface excitation spectra for increasing amounts of adsorbed K. At low K coverages, the Ag SP peak is only weakly affected. As the K coverage approaches one monolayer, the FWHM of the SP increased from 0.17 (clean Ag layers) up to 0.70 eV. Moreover, upon K adsorption the plasma energy of the Ag SP shifted from 3.80 down to 3.56 eV. The redshift of the Ag SP energy may be ascribed to a charge transfer from the Ag substrate to the adsorbates. A reduced Ag SP frequency in the presence of electronegative coadsorbates was reported also for Ag single-crystal surfaces [47, 48]. A major role of potassium should have implied a blue-shift due to the electropositivity of alkali adatoms [49].

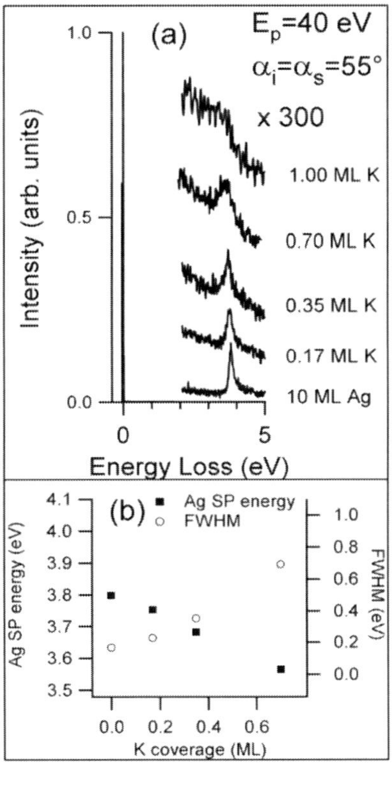

Figure 15. HREEL spectra for K/10 ML Ag/Ni(111) as a function of K exposure. (b) (■) Ag SP energy and (○) FWHM as a function of K coverage.

Vibrational measurements indicated that water contamination occurs. Hence, we suggest that the redshift of the plasmon energy may be water-induced, as found for H_2O/Na/Cu(111) [7]. It is more correct to describe the plasmonic excitation as the K/Ag interface plasmon rather than a redshifted Ag SP. New adsorbate-induced electronic states arose at the interface. Hence, the significant plasmon broadening is due to new channels for decay into electron-hole pairs at the K/Ag interface. The interband transitions involving the overlayer-induced band below the Fermi level thus influences the SP energy and gives rise to a redshift of the SP energy.

Such SP broadening is ascribed to chemical interface damping. It was observed in absorption spectra of small metal particles embedded in a reactive matrix [50, 51], but it is the first experimental evidence of its occurrence in thin films.

Ag/Cu(111)

In Ag/Cu(111) silver islands appear for Ag coverages higher than two layers, which instead grow layer-by-layer. Around each spot of the p(1x1)-Ag, LEED measurements showed also the occurrence of a (9.5x9.5) reconstruction for the first Ag layer and of a (9x9) reconstruction in correspondence of θAg=2.0 ML, in excellent agreement with previous structural studies on this system [52, 53].

Hence, this system offers the opportunity to study the dependence of dispersion and damping dispersion of the SP as a function of the growth mode.

Density-functional theory calculations based on s-d polarization model found an initial negative dispersion for two layers of Ag deposited on Al surfaces [21]. By contrast, experiments [26] carried out on Si(111) reported a positive behavior. It would be extremely useful to study dynamic screening processes in the case of a thickness of two layers. In fact, two layers are commonly accepted as a borderline between interface physics and thin-films physics [21, 54, 55].

As a matter of fact, such coverage constitutes the minimal thickness necessary to observe in the loss spectra well-distinct features assignable to collective excitations, while the very broad loss features observed from 0 to 2 ML are related to single-particle transitions [55].

However, experimental measurements on Ag nanoscale thin films deposited onto metallic substrates which could verify the effects of s-d screening have not yet been performed. On the other hand, the collective excitations in two alkali layers on metal substrates were extensively studied [54-56] as significant differences compared to thick films were found [57]. Contradictory results about the existence of the SP at small momenta were reported [21, 54-56]. The investigation of the behavior of collective excitations also for two Ag layers on metallic substrates should provide a significant advancement in understanding dynamic screening processes at metal surfaces. Moreover, a shifted bulk plasmon (BP) is expected to exist at small momenta, where the SP weight is vanishing [21].

Selected HREEL spectra for 2 ML of Ag as a function of the parallel transfer momentum are showed in Figure 16 (left panel).

In contrast with all previous experimental works on Ag surfaces [22, 37-40] and layer-by-layer Ag films on Si(111) [26], the frequency of the SP did not increase as a function of the parallel transfer momentum (Figure 17).

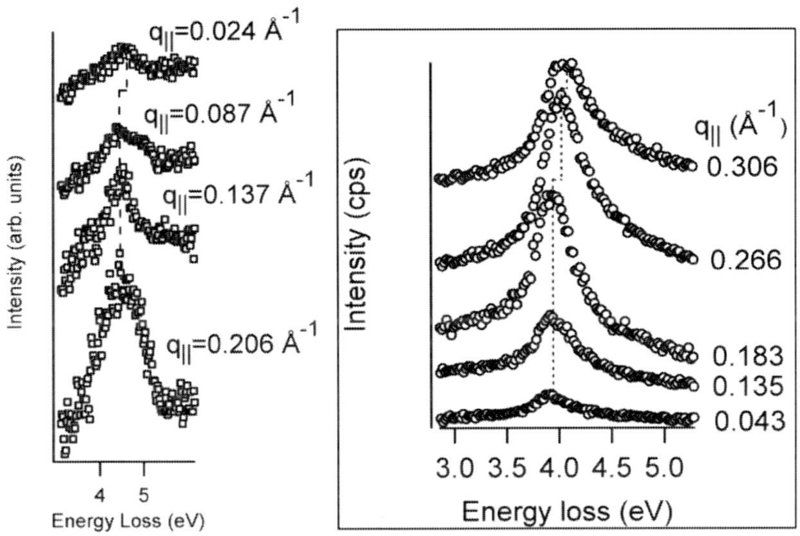

Figure 16. HREEL spectra as a function of the scattering angle for 2 (left) and 5 ML Ag/Cu(111) at T=300 K.

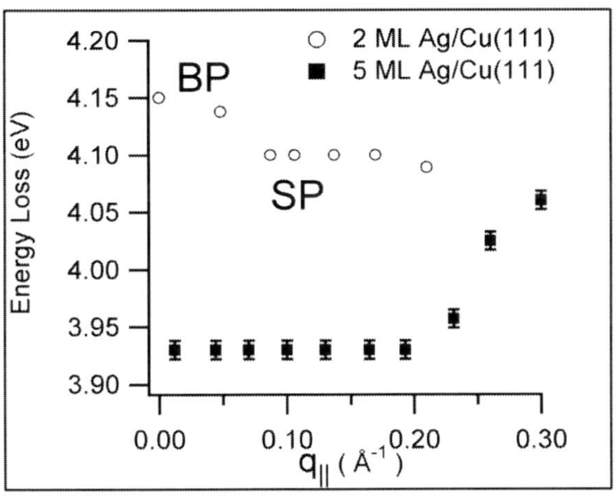

Figure 17. SP dispersion as a function of the parallel transfer momentum for 2 ML and 5 ML of Ag on Cu(111) at T=300 K.

Present results well agree with theoretical calculations by Liebsch [21]. As the thickness of the Ag film is reduced, the overall screening of the charge associated to the SP via the polarizable d electronic medium diminishes, giving higher plasmon energy. Furthermore, the spill-out region not affected by s-d polarization becomes more important, causing a further blue-shift of the plasmon frequency. Moreover, the differences that we found with respect to measurements on 2.5 ML Ag/Si(111) [26] (Figure 2) were ascribed to the enhanced screening properties of metal substrates compared to semiconductor surfaces and to more efficient screening processes due to electron quantum confinement [43].

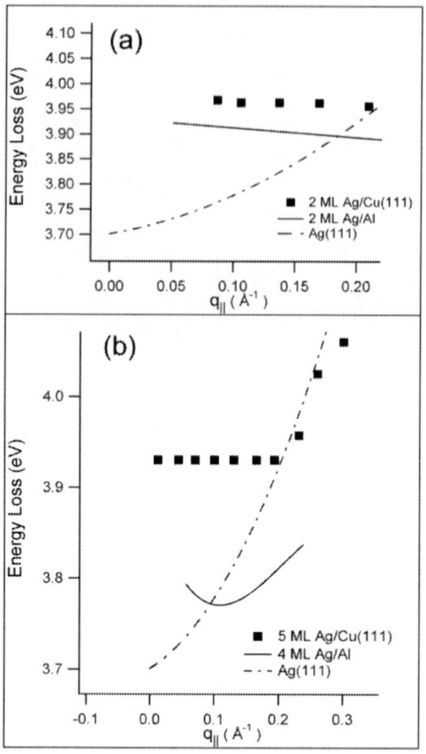

Figure 18. Behavior of the dispersion curve of the Ag SP for Ag(111) (dashed-dotted line) [38], 2 ML Ag/Cu(111) (our data, filled squares), 2 ML Ag/Al (theory, continuous line) [21], and 2.5 ML Ag/Si(111) (dotted line) [26]. The values of the SP energy for the various dispersion curves were normalized to their respective SP energies measured at smallest value of $q_{||}$ in order to put in evidence the different behavior.

Interestingly, for two Ag layers, in the limit of small momenta, we observed the excitation of the BP, shifted by s-d screening at a frequency $\omega_p^*=\omega_p/\sqrt{\varepsilon_d}$, where ε_d is the local dielectric function and ω_p is the s-p BP frequency [21]. The real part of the dielectric function decreases as the film thickness is reduced, as a direct consequence of the occurrence of a less sharp onset of transitions involving d states in thin films compared to bulk Ag [21]. As a matter of fact, a significant energy step between small and high momenta exists due to the different nature of the excitation, i.e. BP and SP, respectively. The lack of the SP excitation at $q_{||}=0$ was reported also for ultrathin alkali layers [56]. On the contrary, in ultrathin Ag layers on Si(111) [20] the SP was excited even at small momenta. Once again, such evidence should be taken as a fingerprint of very different screening processes between Ag/Si and Ag/Cu. However, increasing Ag coverage, i.e. 5 ML (right panel of Figure 16 and Figure 17), the SP was excited also at small values of $q_{||}$, as generally found for thick alkali layers [57] and Ag/Si(111) [26]. In Figure 18 we compare the SP dispersion for different Ag systems. The behavior found for two Ag layers on metal substrates differs from the behavior found in all other Ag systems exhibiting, instead, a positive and quadratic dispersion. As a matter of fact, the dispersion curve calculated by Liebsch [21] for 2 ML Ag/Al well agree with present results for Ag on Cu(111).

The granularity of the as deposited film for coverages higher than two layers is argued from the behaviour of the dispersion of the collective excitation (Figure 19). The absence of dispersion below a critical wave vector, i.e. 0.15 A^{-1}, indicates that the s electrons oscillate independently in the single (111)-oriented grains. Similar results were reported for Ag/Si(111) [24]. The critical wave vector was suggested [24] to be related to the average island size through the relation

$$Qc=2\pi/d$$

From the above equation, we estimate that the grain size in Ag/Cu(111) is about 30-40 Å.

The propagation of the SP can occur only for modes whose wavelength is smaller than the diameter of the single grain. Interestingly, the Ag grains behave like isolated clusters with respect to the plasmonic excitation.

The behaviour of the SP dispersion well agrees with the Stranski-Krastanov growth mode of this system [58-61]. The increasing strain caused by adsorbed layers destabilizes the film and induces clustering [62]. This behavior arises from the large lattice mismatch between Ag and Cu (13%). Low-energy electron diffraction experiments showed the occurrence of a 9x9 overstructure of the deposited silver film, as previously reported [52, 53].

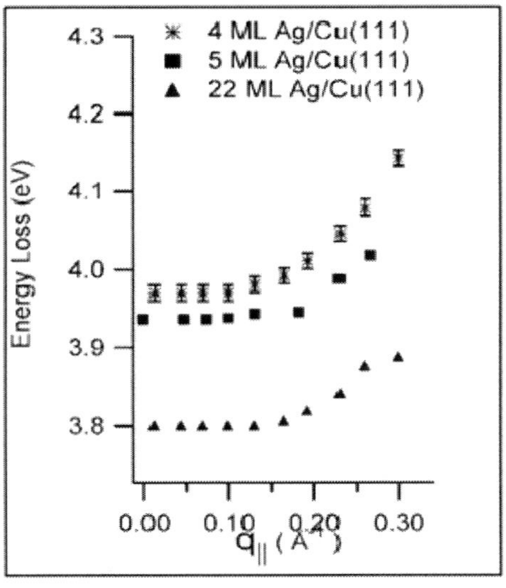

Figure 19. Dispersion of the SP energy for different Ag coverages. The lack of a dispersion before a critical wave-vector indicates that the SP is confined within Ag grains.

It is worth noticing that we have not evidences for the existence of Mie plasmons within Ag islands. Two well-distinct Mie plasmons at 3.1 and 3.9 eV were revealed only for Ag deposited on metal-oxide surfaces [29, 30] and not for three-dimensional islands on semiconductor [24, 26] surfaces. As concerns metal/metal interfaces, the occurrence of Mie plasmons was invoked only for Na quantum dots on Cu(111) [7, 63] but only for a very restricted alkali thickness range. Mie plasmon merged into the ordinary SP already for two nominal Na layers.

In order to remove the SP confinement, an annealing of the film at 400 K was performed. Loss spectra in figure 20 provided evidences for drastic morphological changes in the film and a higher degree of ordering, as suggested by the analysis of the LEED pattern.

Annealing the surface at 400 K caused significant changes in the dispersion curve and, in particular, the loss of the SP confinement. The measured dispersion curve $E_{loss}(q_{\|})$ of the annealed film, reported in Figure 21, was fitted by a second-order polynomial given by:

$$E_{loss}(q_{\|})=A+Bq_{\|}+Cq_{\|}^2$$
(A=3.791±0.006 eV; B=-0.60±0.09 eV·Å; and C=3.4±0.3 eV·Å^2)

The linear coefficient was found to be slightly negative. Recently, it has been found to be null on the similar Ag/Ni(111) system [26]. This finding was ascribed to the enhanced sp density of states existing in thin Ag films, as a direct consequence of the presence of QWS. Increasing the free-electron character of the QWS by annealing [15] should imply the occurrence of a negative linear term.

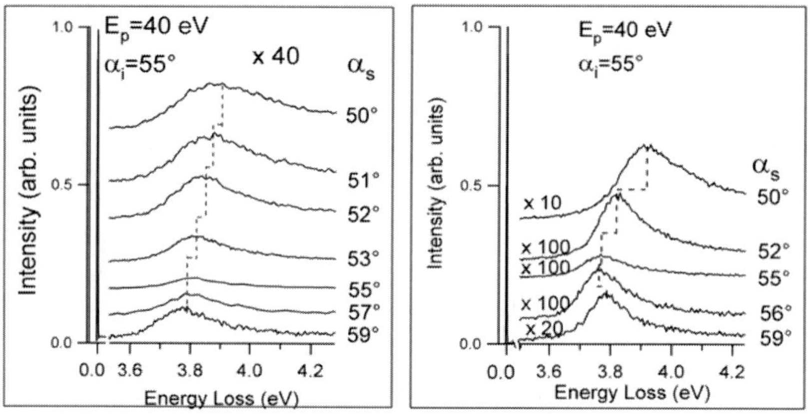

Figure 20. HREEL spectra for as-deposited (left panel) and annealed (right panel) 22 ML Ag/Cu(111).

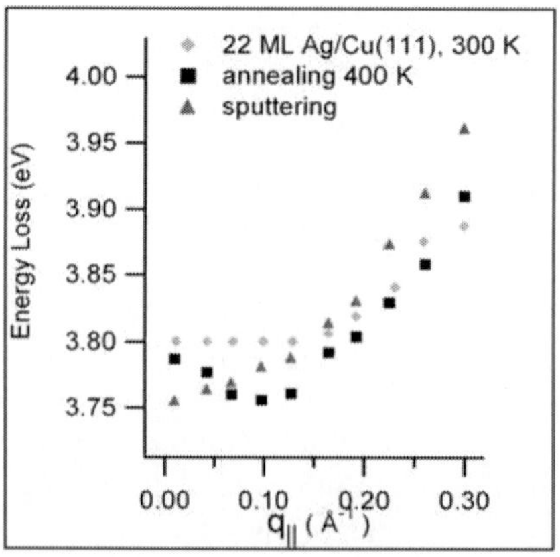

Figure 21. Behavior of the SP dispersion curve.

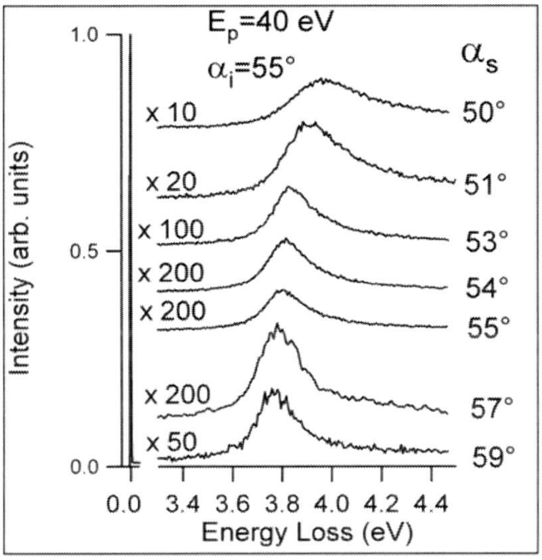

Figure 22. HREEL spectra for a sputtered 22 ML Ag/Cu(111) surface.

Nonetheless, the value of the linear coefficient is still enough higher than the linear coefficient of the SP dispersion curve of alkalis [57], aluminium [64], or alkaline-earth metals [65]. Such finding leads us to suggest that the centroid of the induced charge lies in the close vicinity of the jellium edge [21, 41, 42], in contrast with all other Ag systems [21, 22, 24-26, 37-40], but not outside as for simple metals. Interestingly, the quadratic coefficient coincides with that of SP dispersion in Ag(111) [38], i.e. the surface with the same crystallographic orientation. Significant differences exist between spectra acquired for annealed (figure 20) and sputtered films (figure 22).

The dispersion curve measured in a sputtered Ag film reported in figure 21 shows that the quadratic term is predominant:

(A=3.760±0.004 eV; B=-0.08±0.06 eV·Å; and C=2.5±0.2 eV·$Å^2$).

Accordingly, an increased linear coefficient and a decreased quadratic term were obtained by fit procedure. Contrary to the sputtered Ag(100) surface [66] for the quadratic term the bulk value of 6 eV·$Å^2$ [67] was not recovered. In our opinion, the link proposed in ref. 13 between the value of the quadratic term of the SP dispersion and that of the bulk plasmon, related to bulk properties, is not at all well-founded.

The occurrence of an increased linear coefficient suggests that sputtering induces a significant shift of the position of the centroid of the induced charge associated to the SP compared with that of SP in annealed films.

It is worth mentioning that ion bombardment of a growing film was found to produce both bombardment-induced segregation normal to the film surface and an advancing nanoscale subsurface diffusion zone [68]. Such phenomena should be considered in theoretical studies on the electronic response of sputtered thin films (still lacking). Moreover, our results provide the grounds for angle-resolved photoemission experiments shedding light on the sputtering-induced modifications of the QWS.

Damping Processes of the SP in Ag/Cu(111)

Figure 23 (left panel) shows the thickness-dependence of the full-width at half maximum (FWHM) as a function of the parallel momentum transfer for Ag/Cu(111). An initial negative behavior of the FWHM was found for ultrathin films, while for higher coverages (22 ML) the behavior is substantially dispersionless. A comparison among the dispersion of the SP line-width for 5 ML Ag/Cu(111), 5 ML Ag/Si(111) [26], and Ag(111) [39] shows (Figure 23, right panel) that for Ag films a critical wave-vector beyond which the dispersion became positive exists, while for the single-crystal surfaces the initial dispersion of the FWHM is nearly flat. However, the value of the turning point differs in 5 ML deposited on Cu(111) and Si(111) (0.19 and 0.08 Å^{-1}, respectively).

This behavior of the FHMM is well described by a theoretical model recently proposed [16] on plasmon lifetime in free-standing Ag layers. The behavior of line-width of the Ag SP as a function of the parallel momentum transfer was found to be characterized by a negative behavior of the line-width for small momenta up to a critical wave-vector. This finding was ascribed [46] to the splitting between symmetric and anti-symmetric excitation modes and the enhanced electron-hole pair excitation at small $q_{||}$. For higher values of $q_{||}$, a linear increase of the line-width was reported.

For the as-deposited Ag film, the initial dispersion of the FWHM is negative (Figure 24). The behavior became positive after a critical wave-vector (0.10 Å^{-1}) This finding is in agreement with results for Ag/Si(111) [26] and recent theoretical calculations for free-standing Ag slabs [46]. However, the behavior of the FHWM for single-crystal Ag surfaces is positive [22] and was recovered by annealing the Ag film.

It is worth noticing that in all cases (as-deposited, annealed, and sputtered Ag film), the value of the FWHM is higher than for Ag semi-infinite media (for Ag(111) the FWHM at small momenta is 69 meV [39]). As discussed above, the different distribution of occupied and unoccupied electronic states in thin films with respect to surfaces would imply enhanced damping mechanisms for collective excitations by creating electron-hole pairs.

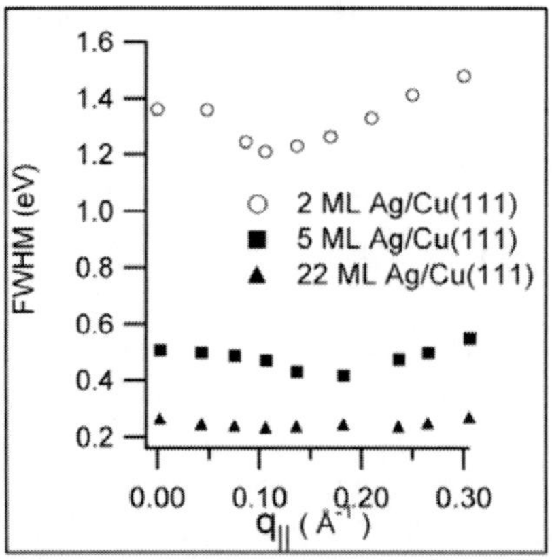

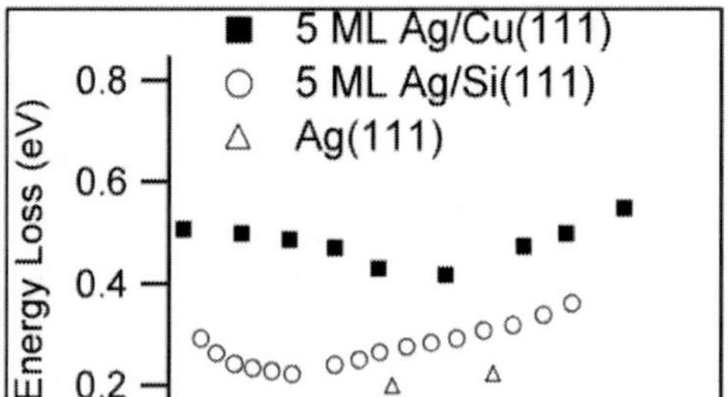

Figure 23. (left) Behavior of the FWHM of the Ag SP as a function of Ag thickness on Cu(111). (right) Comparison between the dispersion of the FWHM of the Ag SP for different Ag system.

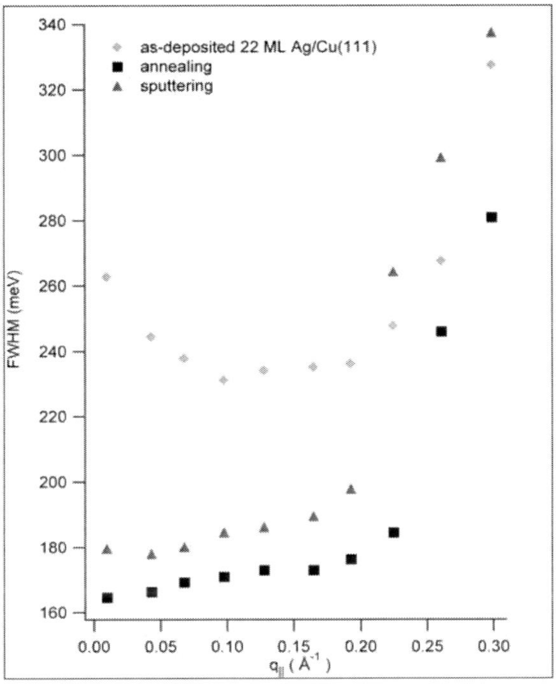

Figure 24. Behavior of the FWHM for as-deposited, annealed, and sputtered Ag films (22 ML) on Cu(111).

The different behavior with respect to the case of Ag/Ni(111) should be ascribed to differences in both the growth mode and in the nature of QWS. In fact, due to the absence of a gap in Ni(111), the character of the quantum well states in such two systems differs substantially [14]. Quantum well states on Ag/Cu(111) are characterized by standing wave patterns, not supported on Ag/Ni(111). Furthermore, on Ag/Cu(111) the interfacial transmittivity is suppressed, with an enhanced specular reflectivity. The opposite occurs for Ag on Ni(111). Such significant differences in the electronic properties between these two bimetallic surfaces should in principle imply quite different Landau damping processes of the plasmonic excitation.

Na/Cu(111)

The (111) surface of noble metals (Cu, Ag, Au) exhibits a large confined gap within the projected bulk band structure centered at the $\overline{\Gamma}$ point of the surface Brillouin zone [70]. The adsorption of Na on Cu(111) has been extensively investigated recently. The existence of Na $3p_z$-derived confined electron states has been well established [71, 72]. Multilayers of alkali metals are easily grown on Cu(111) even at room temperature (RT) while in other systems the growth of a second alkali layer is possible only at liquid nitrogen temperature [73]. A study reporting the dispersion curve of the SP in alkali thin layers at RT is still lacking.

HREEL spectra of Na adsorbed at RT on Cu(111) as a function of coverage are shown in Figure 25. For the clean Cu(111) surface, the onset of collective excitations of valence electrons could be detected at 2.1 eV [74, 75]. A broad loss feature, peaked at 3.70 eV and assigned to the Na SP, gradually arose in the spectrum as a function of Na coverage.

Loss spectra in Figure 26 show the SP of two layers of Na/Cu(111) as a function of the scattering angle for E_p=20 eV. The spectrum recorded in the specular geometry is centered at 3.70 eV and it exhibits clear energy dispersion for off-specular angles.

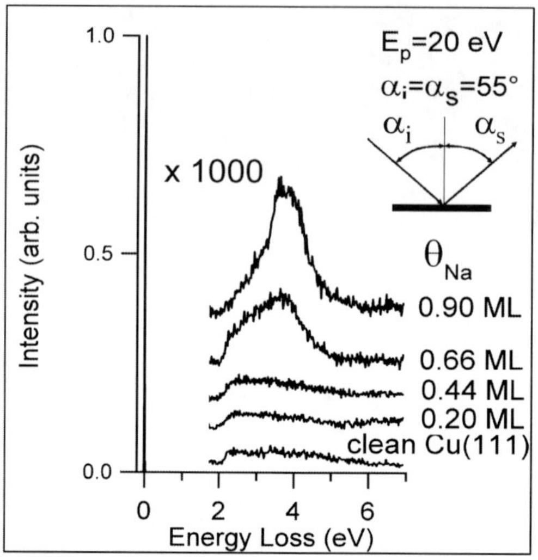

Figure 25. HREEL spectra of Na/Cu(111) as a function of the alkali coverage. The incident beam energy E_p was 20 eV.

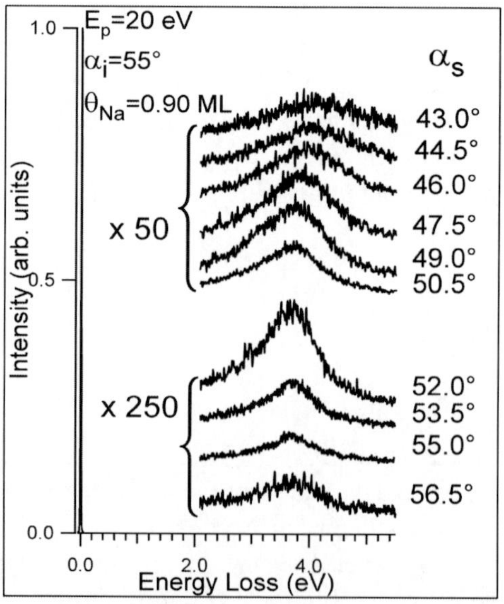

Figure 26. Electron energy loss spectra of 0.90 ML Na/Cu(111) at different scattering angles θ_s. The incident beam energy E_p was held constant at 20 eV and all spectra were recorded at an incident angle of θ_i =55° with respect to the sample normal.

As mentioned above, changing the primary beam energy of the electrons, it is possible to modify their mean free path in the solid and so their penetration length [45]. This allows a direct control of the surface sensitivity. Moreover, the modification of the electron penetration would imply, according to random-phase approximation calculations [76], a change in the position of the reflection plane at which probing electrons are scattered within the extended electron-density distribution. On the basis of the above result, a strong dependence of dynamic screening processes on impinging energy is expected especially for systems in which electrons are confined into a two-dimensional space. Hence, spectra were acquired also for a higher primary energy, i.e. 100 eV, in order to reveal such behavior.

Multipole SP in Na/Cu(111) became quite evident in spectra taken with E_p=100 eV behind 3 degrees off-specular (figure 27). Its energy was found to be 4.70 eV, in excellent agreement with the value reported for bulk Na by Tsuei et al. [77], i.e. 4.67 eV. However, the multipole mode was not revealed for lower impinging electron beam energies, thus suggesting the existence of threshold primary beam energy.

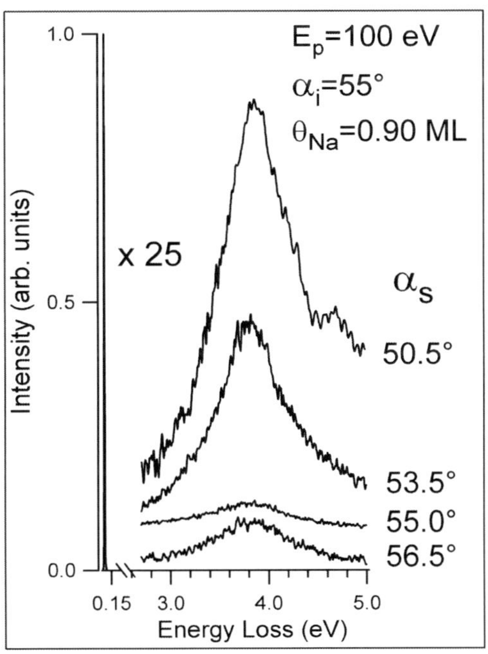

Figure 27. Electron energy loss spectra of 0.90 ML Na/Cu(111) at different scattering angles θ_s, recorded with a primary electron beam energy of 100 eV.

The measured dispersion curve $E_{loss}(q_{\|})$ for E_p=20 eV reported in Figure 28 was fitted by a fourth-order polynomial [22] given by:

$$E_{loss}(q_{\|})=a+bq_{\|}+cq_{\|}^2+dq_{\|}^3+eq_{\|}^4$$

where a=3.99 eV, b=0.70 eV·Å, c=-29.53 eV·Å^2, d=113.14 eV·Å^3, and e=-112.31eV·Å^4.

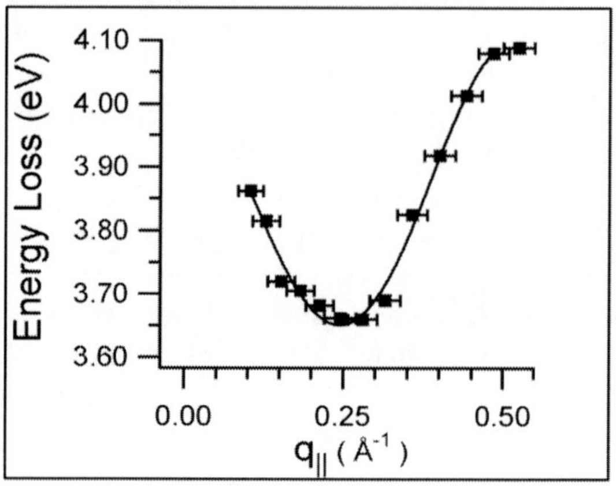

Figure 28. Na SP energy as a function of $q_{\|}$ (E_p=20 eV).

The slope of the dispersion curve is negative up to 0.25 $Å^{-1}$, then the loss energy of the SP increases with increasing $q_{\|}$ and the dispersion becomes definitively positive. It should be noticed that the SP was not damped into single-particle transitions until $q_{\|}$=0.53 $Å^{-1}$, while the critical wave-vector of alkali SPs was found to be around 0.30 $Å^{-1}$ both in thin [55] and thick [57] alkali films.

Furthermore, no loss features were revealed at the lowest values of parallel momentum transfer (below 0.10 $Å^{-1}$). A similar result was found by Zielasek et al. [56] in Cs/Si(111)-(7x7) in which the SP was suppressed for small wave vector. These findings are in agreement with Liebsch's calculations [21] predicting that for two layers of alkali metals the spectral weight of the SP is vanishing at $q_{\|}$=0. In other words, at $q_{\|}$=0 only the excitation of the so called multipole SP is possible. In this regard, we notice that the Na/Cu(111) system behaves differently from K/Ni(111) [55] for which the excitation of the SP was found to be possible.

The energy dispersion of the SP of two Na layers was 430 meV, while it was 130 meV for thick Na films. Such finding could be explained by screening effects that push the position of the induced charge density centroid more outside the substrate jellium edge than in thick Na films.

We propose that electron quantum confinement in Na QWS may be the responsible of such result. In fact, the electron confinement of Na electrons on Cu(111) was found [43] to drastically change the dynamical screening properties of this system. The occurrence of electron confinement in Na QWS, moreover, modifies the electron charge-density distribution. Hence, the reflection of external charges at different distances from the surface [76] should affect the electronic response of this system.

The dispersion curve of the SP, measured using a primary electron beam of 100 eV (figure 29) is quite different from that in Figure 28. The energy dispersion of the SP is lower (70 meV vs. 430 meV). The measured dispersion curve $E_{loss}(q_{\|})$ for E_p=100 eV was fitted by a fourth-order polynomial [22] given by:

$$E_{loss}(q_{\|})=a+bq_{\|}+cq_{\|}^2+dq_{\|}^3+eq_{\|}^4$$

where a=3.84 eV, b=-0.79 eV·Å, c=6.22 eV·$Å^2$, d=-17.98 eV·$Å^3$, and e=-20.99 eV·$Å^4$.

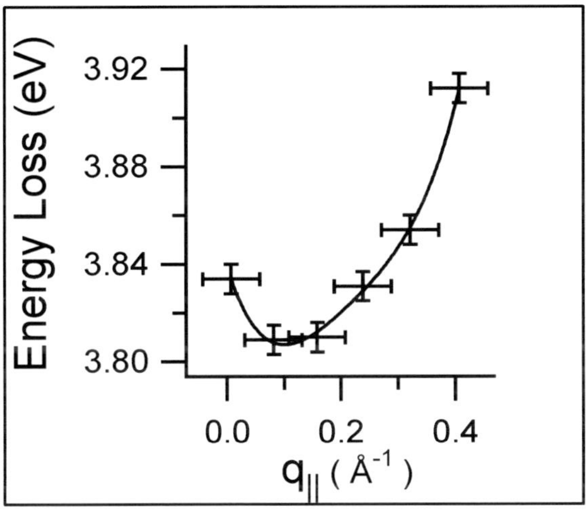

Figure 29. Na SP energy as a function of $q_{||}$ (Ep=100 eV).

It showed an initial negative behavior that became positive above 0.15 $Å^{-1}$, in analogy with the dispersion curve of SPs of other simple metals.

It should be noticed that for E_p=100 eV the SP could be excited even at small momenta. This is a combined effect of the higher dipole/impact scattering ratio of the excitation (see Figure 34 and its discussion) and of the increased integration window [22] in reciprocal space. The latter implies the collection of a finite range of momenta of the outgoing electron.

A dispersion curve not dependent on the primary electron beam energy is expected only for the case of pure sheet plasmons [78], i.e. two-dimensional plasmons strictly confined to the surface and with a square root-like dependence on $q_{||}$. However, HREELS measurements revealed that the dispersion curve depends on the impinging energy. When the thickness of the film gets smaller, the tail of the wave function of the QWS can reach and interact with the substrate. The main effect is a hybridization interaction between overlayer and substrate states. We suggest that the occurrence of such interactions between overlayer states and substrate ones may lead to the observed differences in the two dispersion curves.

Figure 30 shows a comparison between the dispersion curve obtained for two layers of Na on Cu(111) and the dispersion curve of the SP of a thick Na film [57]. The plasmon energy for a nanoscale thin Na film is lower because screening effects make lower the induced average charge-density and thus the SP energy, as found in two layers of K on Ni(111) [55]. The differences in the critical wave-vector of the SP for each dispersion curve are quite evident, thus suggesting strong differences about damping processes.

The damping of the plasmon excitation is clearly revealed by the trend of the full width at half maximum (FWHM) versus $q_{||}$, as shown in Figure 31 for E_p=20 eV. The width of the Na plasmon initially decreased, followed by a steep increase as a function of $q_{||}$. Experimentally, a similar behavior was also observed for a variety of metal surfaces such as Ag surfaces, Mg(0001) [65], Al(111) [64], and Cu(111) [75]. On the contrary, the FWHM of the SP in thick films of Na and K [57] increases for increasing $q_{||}$, while it was almost constant in two layers of K on Ni(111) [55].

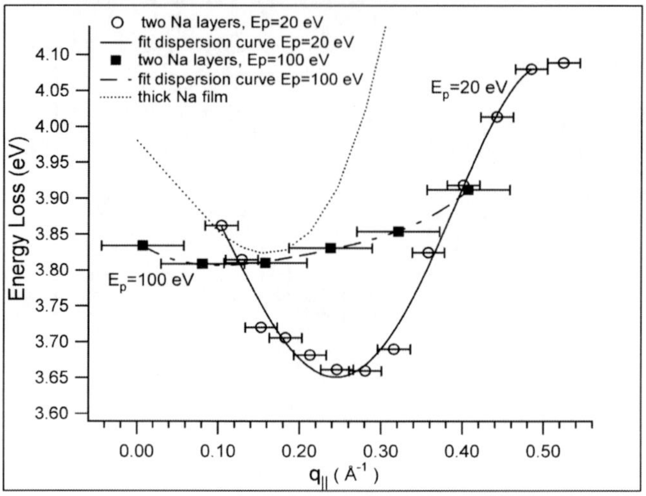

Figure 30. The SP dispersion versus the parallel momentum $q_{||}$ for: (■) two Na layers, E_p=100 eV ; (○) two Na layers, Ep=20 eV ; (-) thick Na film (data taken from ref. 57).

Existing theories [23] predict that, with increasing of the momentum $q_{||}$ parallel to the surface, the width of the SP rapidly increases due to the decay into electron-hole pairs (Landau damping). The behavior shown in Figure 31 could arise from a more efficient disexcitation channel of the SP by single-particle transitions at small momenta. Such assumption could explain a so high critical wave-vector (0.53 $Å^{-1}$) of the plasmonic excitation too.

The FWHM of the SP for E_p=100 eV was characterized by a negative dispersion vs. $q_{||}$ (figure 32). To the best of our knowledge, it is the first time that a FWHM was found to have a similar behavior which disagrees with existing theories. Such finding suggests a strong dependence of the dynamical response of electrons and of screening effects as a function of the impinging energy. Interestingly, the FWHM for thick Na films [57] has the opposite trend as a function of the parallel momentum transfer. This experimental result is a clear evidence that understanding of the broadening mechanisms of a SP requires a careful analysis of the band structure of the system.

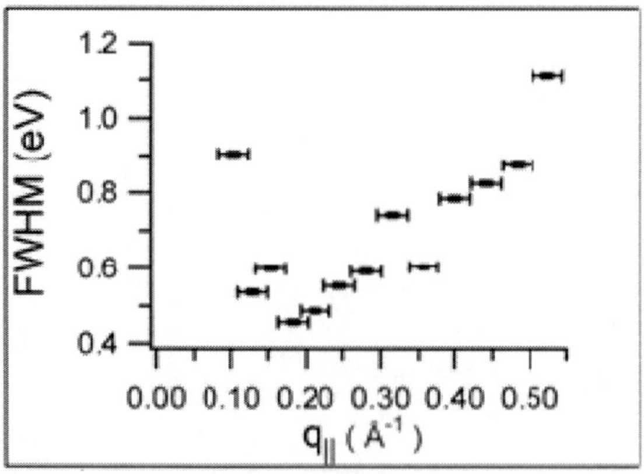

Figure 31. FWHM of the Na SP peak as a function of $q_{||}$ (E_p=20 eV).

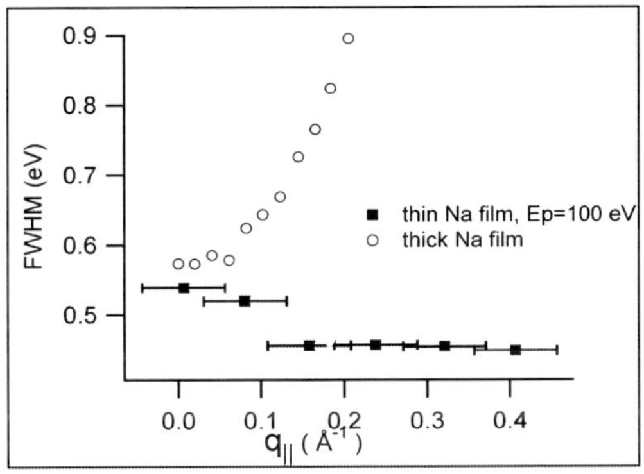

Figure 32. FWHM of the Na SP peak as a function of $q_{\|}$ for (■) 0.90 ML Na/Cu(111), Ep=100 eV and for (○) thick Na film (data taken from ref. 57).

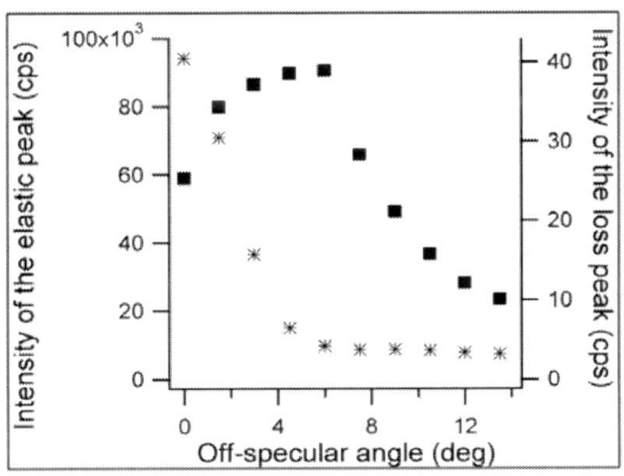

Figure 33. Intensity of the SP (■) and of the elastic peak (*) as a function of the off-specular angle (E_p=20 eV).

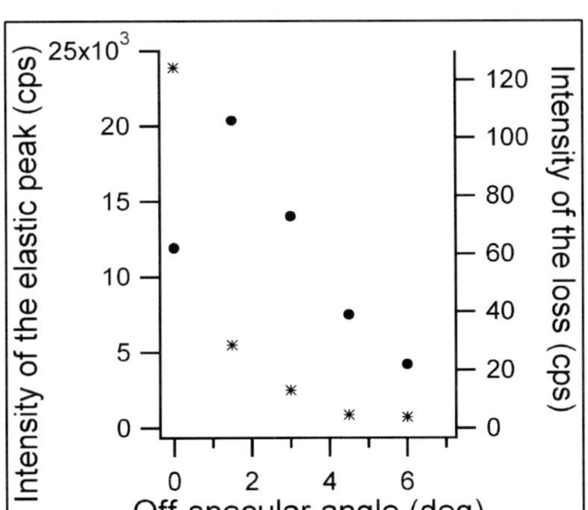

Figure 34. Intensity of the SP (■) and of the elastic peak (*) as a function of the off-specular angle (E_p=100 eV).

The intensity of the SP versus the off-specular angle for E_p=20 eV (Figure 33) clearly demonstrates that such collective mode was excited by impact mechanism because it is peaked at six degrees off-specular [22], while a dipolar loss would have the same behavior of the elastic peak as a function of the off-specular angle. Instead, the intensity of the SP for E_p=100 eV exhibited a maximum around 1.5 degrees off-specular (Figure 34). Thus, the SP excited by a higher primary beam energy has a substantially dipole character while the same mode with a lower impinging energy exhibits a notable impact character. The dependence of the dipole/impact scattering ratio in SP excitation as a function of primary beam energy is still an unexplored research field.

ACKNOWLEDGMENTS

We want prof. Leo Marušić for his help in writing the theoretical background. We want also to thank Dr. Stefan Mathias and Prof. Michael Bauer for having allowed to use their photoemission data, and, moreover, prof. Chiarello and dr. Vincenzo Formoso for many helpful discussions.

CONCLUSIONS

In conclusion, measurements carried out on Ag layer-by-layer films grown on Ni(111) provided evidences of the relationship between the existence on Ag QWS and the linear term of the dispersion curve, which is null in contrast with all other Ag systems.

Similar loss measurements for Ag/Cu(111), characterized by a Stranski-Krastanow growth mode, revealed SP confinement within islands in as-deposited Ag layers. Annealing caused an enhanced free-electron density of states of the Ag QWS, which renders negative the linear coefficient of the dispersion relation.

The increased value of the FWHM in Ag films compared with Ag single-crystal surfaces suggest the occurrence of enhanced damping mechanism due to the opening of new decay channels of the SP in systems exhibiting QWS.

For two layers of Na on Cu(111), the SP was not excited at small momenta, also found for two layers of Ag on the same substrate. A strong dependence of the dispersion curve on the energy of the primary beam was observed.. The FWHM of two Na layers behaves very differently with respect to that reported for a Na thick film. These differences were ascribed to different Landau damping mechanisms of the plasmonic mode in the two cases.

Screening effects enhanced by electron quantum confinement and interactions between overlayer states and substrate states are suggested to be at the basis of these results. Such findings provide the grounds for theoretical studies aimed at characterizing of the nature and dispersion of the excitation modes in nanoscale thin films exhibiting confined electron states.

REFERENCES

[1] Wei, C. M.; Chou, M.Y. *Phys. Rev. B* 2002, *66*, 233408.
[2] Valla, T.; Kralj, M.; Siber, A.; Milun, M.; Pervan, P.; Johnson, P. D.; Woodruff, D. P. *J. Phys. Condens. Matter* 2000, *12*, L477.
[3] Luh, D.-A.; Miller, T.; Paggel, J. J.; Chiang, T.-C. *Phys. Rev. Lett.* 2002, *88*, 256802.
[4] Pfennigstorf, O.; Petkova, A.; Guenter, H. L.; Henzler, M. *Phys. Rev. B* 2002, *65*, 045412.
[5] Orr, G.; Jaeger, H. M.; Goldman, A. M. *Phys. Rev. Lett.* 1984, *53*, 2046.
[6] Chiang, T.-C. *Science* 2004, *306*, 1900.
[7] Politano, A.; Agostino, R. G.; Colavita, E.; Formoso, V.; Chiarello, G. *J. Chem. Phys.* 2007, *126*, 244712.
[8] Kevan, S. D. *Phys. Rev. Lett.* 1983, *50*, 526.
[9] Carlsson, J. M.; Hellsing, B. *Phys. Rev. B* 2000, *61*, 13973.
[10] Dil, J. H.; Kim, J. W.; Gokhale, S.;Tallarida, M.; Horn, K. *Phys. Rev. B* 2004, *70*, 045405.
[11] Yu, Y.; Jiang, Y.; Tang, Z.; Guo, Q.; Jia, J.; Xue, Q.; Wu, K.; Wang, E. *Phys. Rev. B* 2005, *72*, 205405.
[12] Milun, M. ; Pervan, P. ; Woodruff, D. P. *Rep. Prog. Phys.* 2002, *65*, 99.
[13] Mathias, S.; Wiesenmayer, M.; Aeschlimann, M.; Bauer, M. *Phys. Rev. Lett.* 2006 *97*, 236809.
[14] Mathias, S.; Wessendorf, M.; Passlack, S.; Aeschlimann, M.; Bauer, M. *Appl. Phys. A* 2006, *82*, 439.
[15] Luh, D.-A.; Cheng, C.-M.; Tsai, C.-T.; Tsuei, K.-D.; Tang, J.-M. *Phys. Rev. Lett.* 2008, *100*, 027603.
[16] Newns, D. M. *Phys. Rev. B* 1970, *1*, 3304.
[17] Marušić, L.; Šunjić, M. *Phys. Scrip.* 2001, *63*, 336.
[18] Despoja, V.; Marušić, L.; Šunjić, M., *Fizika A* 2005, *14*, 207.
[19] Marušić, L.; Despoja, V.; Šunjić, M. *J. Phys.: Condens. Matter* 2006, *18* 4253-4263.
[20] Despoja, V.; Marušić, L.; Šunjić, M. *Sol. Stat. Comm.* 2006, *140*, 270-275.
[21] Liebsch, A. *Electronic Excitations on Metal Surfaces* Plenum, New York, US 1997; Chapter 3.
[22] Rocca, M. *Surf. Sci. Rep.* 1995, *22*, 1.
[23] Pitarke, J. M.; Silkin, V. M.; Chulkov, E. V.; Echenique, P. M. *Rep. Prog. Phys.* 2007, *70*, 1.
[24] Politano, A.; Agostino, R. G.; Colavita, E.; Formoso, V.; Chiarello, G. *Phys. Stat. Sol. RRL* 2008, *2*, 86.
[25] Moresco, F.; Rocca, M.; Hildebrandt, T.; Henzler, M. *Phys. Rev. Lett.* 1999 *83*, 2238.
[26] Yu, Y.; Jiang, Y.; Tang, Z.; Guo, Q.; Jia, J.; Xue, Q.; Wu, K.; Wang, E. *Phys. Rev. B* 2005, *72*, 205405.
[27] Fujikawa, Y.; Sakurai, T.; Tromp, R.M. *Phys. Rev. Lett.* 2008, *100*, 126803.
[28] Chelarua L. I., Meyer zu Heringdorf F. J. *Surf. Sci.* 2007, *601*, 4541.
[29] Lazzari, R.; Jupille, J.; Layet, J.-M. *Phys. Rev. B* 2003, *68*, 045428.
[30] Nilius, N.; Ernst, N.; Freund, H. J.; *Phys. Rev. Lett.* 2000, *84*, 3994.

[31] Varykhalov, A.; Shikin, A.M.; Gudat, W.; Moras, P.; Grazioli, C.; Carbone, C.; Rader, O. *Phys. Rev. Lett.* 2005, *95*, 247601.
[32] Mróz, S.; Jankowski, Z. *Surf. Sci.* 1995, *322* 133.
[33] Nakanishi, S.; Umezawa, K.; Yoshimura, M.; Ueda, K. *Phys. Rev. B* 2000, *62* 13136.
[34] Mróz, S.; Jankowski, Z.; Nowicki, M. *Surf. Sci.* 2000, *454-456*, 702.
[35] Itoh, T.; Umezawa, K.; Nakanishi, S. *Appl. Surf. Sci.* 1998, *130-132*, 497.
[36] Besenbacher, F. ; Nielsen, L. P.; Sprunger, P.T. in *The Chemical Physics of Solid Surfaces*, edited by King, D. A. ; Woodruff, D. P. (Elsevier, Amsterdam,197) vol. 8 p. 237.
[37] Suto, S.; Tsuei, K.-D.; Plummer, E. W.; Burstein, E. *Phys. Rev. Lett.* 1989, *63*, 2590; Rocca, M.;, Lazzarino, M.; Valbusa, U. ibid. 1991, *67*, 3197; Lee, G.; Sprunger, P.T.; Plummer, E. W.; Suto, S. ibid. 1991, *67*, 3198.
[38] Moresco, F., Rocca, M.;, V. Zielasek, T. Hildebrandt, M. Henzler, *Surf. Sci.* 1997, *338*, 24.
[39] Rocca, M.; Lazzarino, M.; Valbusa, U. *Phys. Rev. Lett.* 1992, *69*, 2122.
[40] Rocca, M.; Valbusa, U. *Surf. Sci.* 1993, *287/288*, 770.
[41] Feibelman, P. J. *Prog. Surf. Sci.* 1982, *12*, 287.
[42] Feibelman, P. J. *Surf. Sci.* 1993, *282*, 129; *Phys. Rev. Lett.* 1994, *72*, 788.
[43] Silkin, V.M.; Quijada, M.; Diez Muino, R.; Chulkov, E.V.; Echenique, P. M. *Surf. Sci.* 2007, *601*, 4546.
[44] Tang, S.-J.; Miller, T.; Chiang, T.-C. *Phys. Rev. Lett.* 2006, *96*, 038602.
[45] De Crescenzi, M.; Piancastelli, M. N. *Electron Scattering and Related Spectroscopies*, World Scientific, Singapore 1996.
[46] Yuan, Z. ; Gao, S. *Surf. Sci.* 2008, *602* (2008).
[47] Savio, L.; Vattuone, L.; Rocca, M. *Phys. Rev. B* 2000, *61* 7324.
[48] Kim, J.S.; Chen, L.; Kesmodel, L.L.; Garcia-Gonzalez, P. ; Liebsch, A. *Phys. Rev. B* 56 (1997) R4402.
[49] Moresco, F.; Rocca, M.; Hildebrandt, T. ; Zielasek, V.; Henzler, M. *Surf. Sci.* 1999, *424* 62.
[50] Persson, N. J. *Surf. Sci.* 1993, *281* 153.
[51] Hoevel, H.; Fritz, S.; Hilger, A.; Kreibig, U.; Vollmer, M. *Phys. Rev. B* 1993, *48* 18178.
[52] Bendounan, A.; Forster, F.; Ziroff, J., Schmitt, F.; Reinert, F. *Phys. Rev. B* 2005, *72*, 075407.
[53] Schiller, F.; Cordon, J.; Vyalikh, D.; Rubio, A.; Ortega, J.E. *Phys. Rev. Lett.* 2005, *94*, 016103.
[54] Tsuei, K.-D.; Heskett, D.; Baddorf, A. P.; Plummer, E. W. *J. Vac. Sci. Technol. A* 1991, *9*, 1761.
[55] Chiarello, G. ; Cupolillo, A.; Amoddeo, A.; Caputi, L. S.; Papagno, L.; Colavita, E. *Phys. Rev. B* 1997, *55*, 1376.
[56] Zielasek, V.; Rönitz, N.; Henzler, M.; Pfnür, H. *Phys. Rev. Lett.* 2006, *96*, 196801.
[57] Tsuei, K.-D.; Plummer, E. W.; Feibelman, P. J. *Phys. Rev. Lett.* 1989, *63*, 2256.
[58] Wessendorf, M.; Wiemann, C.; Bauer, M.; Aeschlimann, M.; Brune, H.; Kern B. *App. Phys. A* 2004, *78*, 183.
[59] Bendounan, A.; Fagot-Revurat, Y.; Kierren, B.; Bertran, F.; Yurov, V.Y.; Malterre, D. *Surf. Sci. Lett.* 2002, *496*, L43.
[60] Mueller, M.A.; Samsavar, A.; Miller, T.; Chiang, T.-C. *Phys. Rev. B* 1989, *40*, 5845.

[61] Mathias, S.; Wessendorf, M.; Passlack, S.; Aeschlimann, M.; Bauer, M. *App. Phys. A* 2005, *82*, 439.
[62] Tu K. N., Mayer J. V., Feldman L.C. *Electronic Thin Film Science* 1992 Macmillan, New York.
[63] Politano, A.; Agostino, R. G.; Colavita, E.; Formoso, V.; Chiarello, G. *Surf. Sci.* 2007, *601*, 2656.
[64] Chiarello, G.; Formoso, V.; Santaniello, A.; Colavita, E.; Papagno, L. *Phys. Rev. B* 2000, *62*, 12676.
[65] Sprunger, P.T.; Watson, G.M.; Plummer, E.W. *Surf. Sci.* 1992, *269-270*, 551.
[66] Savio, L.; Vattuone, L.; Rocca, M. *Phys. Rev. B* 2003, *67*, 045406.
[67] Zacharias, P.; Kliever, K.L. *Solid State Commun.* 1976, *18*, 23.
[68] He, J.H.; Carosella, C.A.; Hubler, G.K.; Qadri, S.B.; Sprague, J.A. *Phys. Rev. Lett.* 2006, *96*, 056105.
[69] Chiang, T.-C. Surf. Sci. Rep. 2000, *39*, 181.
[70] Lindberg, V.; Hellsing, B. *J. Phys. Cond. Mat.* 2005, *17*, S1075.
[71] Carlsson, J. M.;Hellsing, B. *Phys. Rev. B* 2000, *61*, 13973.
[72] Hoffmann, G. ; Berndt, R.; Johansson, P. *Phys. Rev. Lett.* 2003, *90*, 46803.
[73] Bonzel, H.P. ; Bradshaw, A.M.; Ertl, G. ed. *Alkali Adsorption on Metal and Semiconductors*, Elsevier, Amsterdam, (1989).
[74] *Handbook of Optical Costants of Solids*, ed. by E.D. Palik (Academic Press, New York, (1985).
[75] Politano, A.; Chiarello, G.; Formoso, V.; Agostino, R. G.; Colavita, E.; *Phys. Rev. B* 2006, *74*, 081401(R).
[76] Nazarov, V. U. *Phys. Rev. B* 1999, *59*, 9866.
[77] Tsuei, K. D. ; Plummer, E. W. ; Liebsch, A. ; Kempa, K.; Bakshi, P. *Phys. Rev. Lett.* 1990, *64*, 44.
[78] Nagao, T.; Hildebrandt, T.; Henzler, M.; Hasegawa, S. *Phys. Rev. Lett.* 2001, *86*, 5747; *Surf. Sci.* 2001, *493*, 680.

In: Plasmons: Structure, Properties and Applications
Editors: Anton E. Turunen and Johannes O. Niemi
ISBN: 978-1-62100-060-0

Chapter 4

GENERATION OF SURFACE PLASMON RESONANCE FOR SUB-WAVELENGTH NANO-PHOTOLITHOGRAPHY

Scott Lewis and Lucio Piccirillo
The University of Manchester, UK

ABSTRACT

A novel photolithographic technique using a periodic hexagonal close packed silver nanoparticle 2D array photo mask has been demonstrated to transfer a nano-pattern into a photoresist using I line proximity photolithography. A 30 nm hexagonal close packed silver nanoparticle 2D array pattern with a 50 nm period has been successfully transferred into S1813 photoresist using I-line exposure wavelength. The resultant feature sizes were 34 nm with a period of 46 nm, due to the surface plasmon resonance where the S1813 photoresist feature is approximately 11 times smaller than I-line exposure wavelength.

1. INTRODUCTION

In 1958, Jack Kilby and Robert Noyce invented the integrated circuit (IC). Electrical circuits could perform logical functions of great complexity by the high integration of active and passive devices such as transistors, resistors and capacitors in a semiconductor wafer. In 1965 it was observed by Gordon E. Moore that the number of integrated devices (transistors) approximately doubled every two years, which has been known as 'Moore's Law' [1]. This law has driven the semiconductor industry to maintain this observation by the miniaturization of devices for the increased device density per unit area. Figure 1 shows the relationship of critical device dimensions and transistor density in the Intel microprocessor [2].

The transistor density has increased by approximately five orders of magnitude where as the physical dimensions of the devices have been reduced from approximately 20μm to the deep sub-μm regime. Furthermore, in the past years the total manufacturing cost of an integrated circuit has essentially remained constant as the next generation devices have adopted similar production methods.

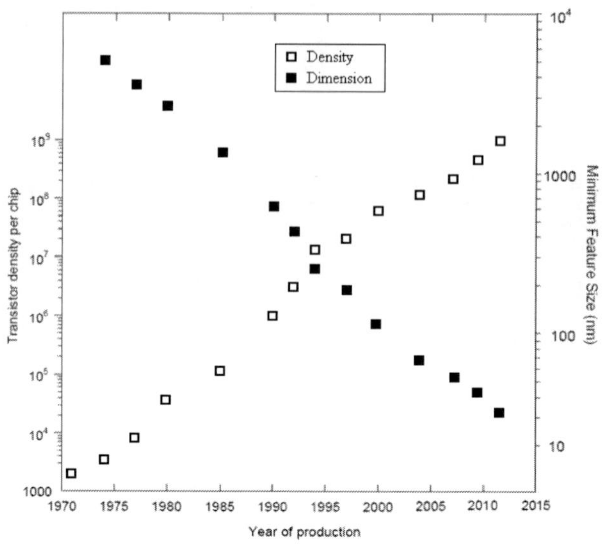

Figure 1. The relationship between the minimum feature size and the density of transistors in an integrated circuit from [2].

Since the early 1970's microprocessors have progressed from 4-bit integrated circuits to 64-bit integrated circuits found in computers today. Not only have microprocessors made a leap forward but memory has also advanced from a capacity of approximately 1 Kbit in the early 1970's to the 100's Gbit dynamic random access memories (DRAM) seen today.

The main component of an integrated circuit is a metal oxide semiconductor field effect transistor (MOSFET). The direct impact of the tremendous miniaturization of MOSFETs is that their performance is enhanced. This is due to the scaling factors associated with the physical dimensions of the device, substrate doping and swing voltages. These factors directly enhance the speed, reduce power dissipation and give a reduction in the power delay product [3]. As these scaling factors enhance the overall performance of the integrated circuit, the cost per functionality is decreased as more and more MOSFET's are manufactured into a single integrated circuit. The structure of the MOSFET seen in Figure 2 has not been modified but the dimensions of the MOSFET have decreased. This has been achieved by reducing the physical dimensions of the width and length.

Micro and nano lithography have played a dominant role in decreasing the dimensions of MOSFET devices, this has been made possible by photolithography. Since the 1970's, lithography has progressed from the micron scale into the nanometre scale.

The economics of the semiconductor industry is the sole driving force behind the improvement in lithography and Moore's Law. This entails that the performance is improved while the costs associated to lithography must remain constant. Lithography is possibly the most critical of the processing steps as about half of the capital equipment cost for a wafer fab is in lithography.

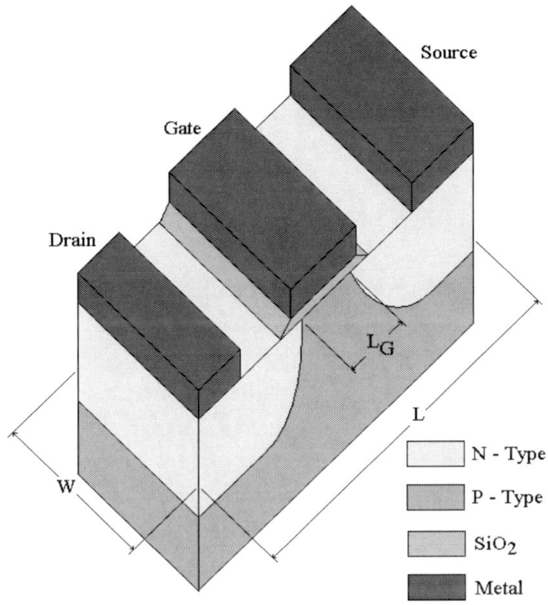

Figure 2. Schematic of a MOSFET structure.

For calculating the lithography costs many models exist. The critical issues of the lithography process can be calculated by this approximation given by equation 1 – 0 [4].

$$Cost/Level = \frac{Tool}{Throughput} + \frac{Mask}{Usage} + Process \qquad (1\text{-}0)$$

When calculating the cost for printing a circuit level on a wafer, there are three main providers. These are the costs that are associated to the method of exposure, mask, and the processing costs.

The first term approximates the relation of the tool costs are obtained by taking the time that the exposure tool is used (labour, depreciation, space, etc.) and dividing it by the time taken to print those wafers – the throughput. This has motivated equipment manufacturers to produce tools with high throughput as this has been the main factor in lithography costs.

The second term in the equation considers the cost of the mask by the number of wafers to be printed with the mask. Approximately 1500 masks are needed, when fabricating a microprocessor as the next generation of microprocessor dimensions are decreased and improved. The current trend seen nowadays is that each mask is used less and less as the cost of the mask increases; this is due to the rapid design revisions [4]. It has been seen that Mask Error Enhancement Factor (MEEF) becomes larger, as minimum dimension of the structure produced is less than the exposure wavelength. Mask enhancements are required such as Phase Shift Masks (PSM) [5] or Optical Proximity effect Correction (OPC) [6].

The third term in the equation determines the estimated process costs this includes the cost of the resist development and its purpose. These days, the process cost is small compared to the tool and mask terms, because it is based on a single – layer resist process.

1.1. The Lithographic Process

For the production of semiconductor devices, patterning is required by selecting areas so that subsequent microelectronics processing steps can be employed, such as etching, deposition, diffusion, or ion implantation for doping. This technique of producing patterns is referred to as 'lithography' which physically 'masks off' the desired and exposes undesired areas for future processing. Figure 3 depicts the fundamental steps that make up the lithography process.

The aim of the lithographic process is to protect selected areas of the device that are not to be modified during a particular processing step like an etching process, by coating that area. The criteria of the lithographic process is that it must be able to protect a selected area of a device by creating a pattern of the desired feature size by fabricating a material referred to as a 'photoresist'. The photoresist is 'photo' sensitive and it 'resists' chemical and physical attack of the underlying substrate. The photoresist material is an organic polymer that consists of dissolution inhibitors and photoacid generators. The dissolution inhibitors are extended chains of molecules that are not dissolved in a developer. When striked by a photon, the photoacid generators produce a photoacid. The photoresist during the different processing steps is illustrated by Figure 4.

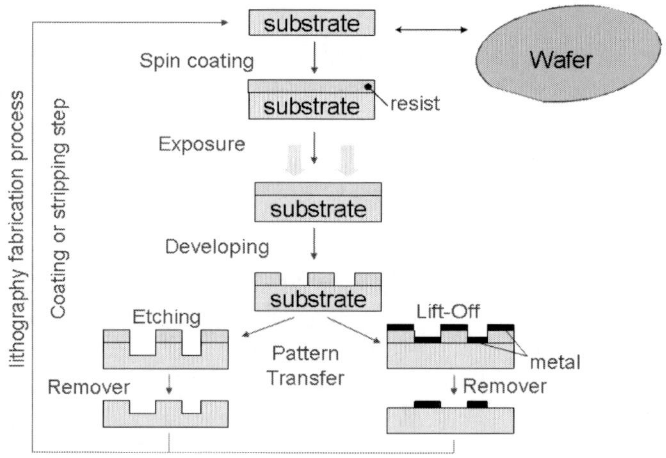

Figure 3. Schematic of the lithographic process.

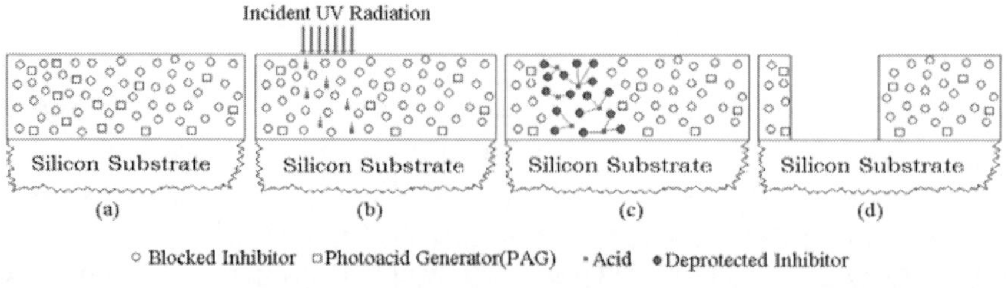

Figure 4. Schematic presentation of the functionality of a CAR. (a) the photoresist before exposure, (b) radiation induced generation of photoacid in exposed areas, (c) acid catalyzed thermal deprotection of inhibitor during PEB and (d) chemical development of the photoresist.

The first step in the lithography process is that the contaminants on the surface of Silicon must be removed by using cleaning chemicals, which may result in poor adhesion and form defects in the photoresist film. However, an adhesion promoter (such as HMDS) improves the adhesion between the photoresist and the substrate that the device will be made from. To produce a uniform coating of the photoresist on to a semiconductor substrate, it is spun from a solution that contains the polymer that is dissolved in a cast solvent.

The semiconductor substrate coated with a photoresist is baked on hot plate at a temperature beyond the boiling point of the dissolving solvent. This is known as a post bake (PB). This is done to evaporate the cast solvent from the photoresist film and to thermally anneal the residual stress that has occurred during spin coating. Figure 4a demonstrates that both the dissolution inhibitors and the photoacid generators (PAG) are uniformly distributed within the photoresist.

The pattern is produced using a Chrome mask. This contains the information that is to be transferred into a polymer in contact with the photoresist. The photoresist is irradiated with ultra violet (UV) radiation. This causes the exposed areas of the photoresist to undergo a chemical transformation by altering the solubility of these areas of the photoresist in the developing stage. Figure 4b depicts the interaction of incident light on the central part of the resist and initializes the photoacid generators (PAG) to produce an acid. The acid catalyzed deprotection of inhibitor molecules is triggered from the supply of thermal energy during the PEB, which can be seen from figure 4c.

The solubility of the resist during the development process is deprotected by the long chains of the inhibitor molecules. It should be noted that in such a reaction the acid is not consumed. One acid molecule may add to the contribution to the deprotection of several inhibitor molecules.

The complexity of these processes in chemical amplified resist (CAR) systems is more complicated. To control the concentration of acid that is generated in certain parts of the resist, recent CARs also incorporate a quencher base into the fabrication of the resist. The mobility of the acid is increased when the source of thermal energy is applied to the deprotection of inhibitor resulting in complex diffusion phenomena [7].

Figure 4d shows the concluding step in the processing sequence. This is the chemical development of the resist. The areas of the resist with a decreased concentration of inhibitor are dissolved.

The photoresist can have tone which is either positive or negative. The difference between these is that if the exposed areas are rinsed away when developed with a solvent it is said to have a positive tone or if the exposed areas remain it is negative tone. The developing step of the lithographic process transforms the image that was irradiated with ultra violet radiation into a three dimensional structure. The structures produced in the photoresist can be removed using a solvent called 'Acetone' or a dry etching technique that incorporates oxygen plasma without damaging the device structure underneath the photoresist.

1.2. Next Generation Lithography Technologies

Figure 5 shows the historic improvements in IC lithography resolution. This has been driven by decreasing the wavelength of the illumination sources. The progression from G –

line at 436nm to I – line at 365nm is illustrated by the figure. Mercury arc lamps filtered for different spectral lines were used as the illumination sources.

This was followed by a conversion to KrF at 248nm by the introduction of excimer laser sources, and in recent times, ArF at 193nm was introduced [4]. Before the establishment of KrF lithography, the minimum structures produced in practice were 250nm as this was larger than the wavelength. However, developments was made in 1999, when the 180nm technology was introduced as this was achieved when structures produced that were printed considerably below the wavelength (248nm) [4].

Production of features with geometry significantly smaller than the wavelength of the exposure radiation, where it depicts that the minimum feature size achievable becomes smaller than the exposure radiation, is known as sub wavelength lithography. This can be largely attributed to the improvements in the system optics and resist materials. With the blend of high – contrast imaging material and good process (exposure dose) control can reliably produce sub wavelength structures.

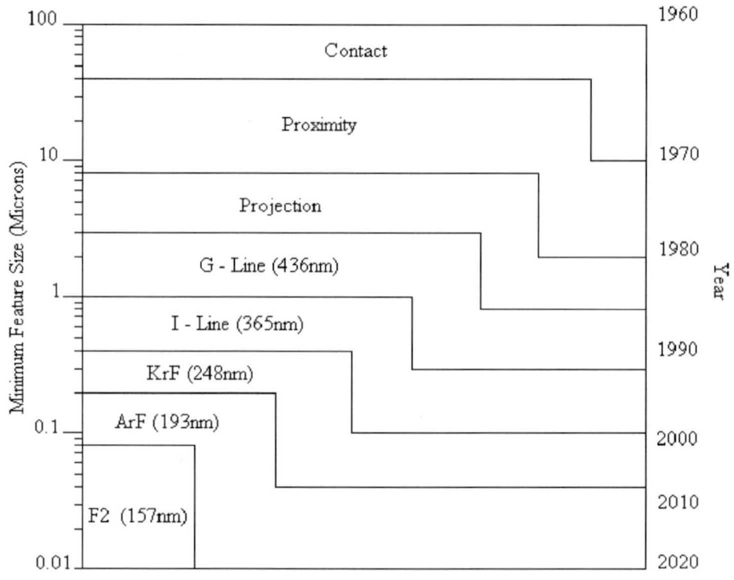

Figure 5. Trends of lithography technologies and exposure wavelengths from [8].

Even though the light intensity image has below full modulation for the nano structures, the best lithographic performance is seen at 193nm. However, unless the acid generators in the photoresist materials for the shorter wavelengths such as 157nm (F_2 excimer) and beyond can be developed to a performance point equal to or better than that for 248nm materials, continued feature size shrinkage through wavelength reduction is not feasible.

Therefore, as seen in Figure 5 the limits of photolithography draws nearer and nearer. New technologies must be developed to enable the transition from the dimensions of hundreds of nanometers to tens of nanometers; via 'NGL' or 'Next Generation Lithography'. The operation of any of the NGL technologies has been far later than predicted as the optical lithography techniques continue to provide the imaging needed at a justifiable cost. This has surprised many and dashed the expectations of several technology ventures. However, for decades the industry's fascination with any NGL technology referring to it as 'the technology of the future', as the forecast to the end of the optical lithography era is getting closer, thus dawning the NGL era.

Eventually, the capability of optical lithography reaches the end as the laws of physics triumph. The question will be 'could the other non-lithography related areas of the industry (such as design, materials, processing methods etc) further advance the nanoelectronics products. If this is so, then it may prevent the need to employ any of the NGL technologies, thus further extending the usefulness of optical lithography. Or, perhaps optical lithography will no longer be the limiting factor as the non-lithography – related nanoelectronics manufacturing factors will reach their own fundamental limits before optical lithography does. The assessment of the demise of optical lithography will be a technology versus financial decision. Regardless of the timing or business case, the challenges facing NGL techniques are enormous.

The main NGL technologies are Extreme ultra violet (EUV), Electron beam lithography (EBL), Proximity x – ray lithography (PXL) and Ion beam lithography (IBL) [8]. These NGL technologies employ significantly different types of sources from 13.5nm photons, 100KeV electrons, ~1nm photons and 75KeV ions respectively, for patterning wafers. The distinction between these technologies and optical technologies are the differences in mask materials and their overall architectures. New pattern transfer challenges must be faced as the exposure wavelengths are decreased. The transmission of the radiation requires equipment that is necessary to handle a vacuum for all the NGL methods. This will become more expensive and the high throughput that is seen for optical lithography will inevitably be decreased.

Therefore, photolithography in the optical near field has been developed as a promising parallel route to the fabrication of nano features beyond the diffraction limit. To perform near-field photolithography several methods have been demonstrated which include the probe of a near-field scanning optical microscope, contact imaging through a transmission mask or a binary phase-shift mask [9,10], evanescent near-field lithography with an embedded-amplitude mask [11], light coupling masks [12], plasmon printing and evanescent interferrometric methods [13, 14] have been successfully demonstrated for performing near – field photolithography.

Surface plasmon resonance (SPR) has the potential to influence light by enhancing sub-wavelength structures. The scientific community has become interested in the phenomenon of the transmission of opaque metal films when light is enhanced through sub-wavelength hole arrays [15, 16]. The excitation of SPRs on the metal surface dramatically enhances the optical throughput via the sub-wavelength apertures. The surface plasmon dispersion relation, ksp which incorporates the frequency-dependent SP wave vector, describes the interaction of light and SPR [17], this is shown here

$$\kappa_{SP} = \kappa_0 \sqrt{\frac{\varepsilon_d \varepsilon_m}{\varepsilon_d + \varepsilon_m}} \qquad (1)$$

where κ_0 is the wave vector of light in vacuum, ε_m and ε_d are relative dielectric constants of a metal and the surrounding dielectric material. The SPR dispersion curve, shown in Figure 6a, indicates that smaller features are expected in the lithographic pattern. The wavelength of the excited SPs is shorter than the wavelength of the excitation light at the same frequency. Due to the momentum mismatch between the light waves and the SP waves, light cannot excite SPs in the metal directly. However, by using a rough surface, a grating coupler, or an attenuated total reflection (ATR) coupler, light can be converted into SPs and vice versa.

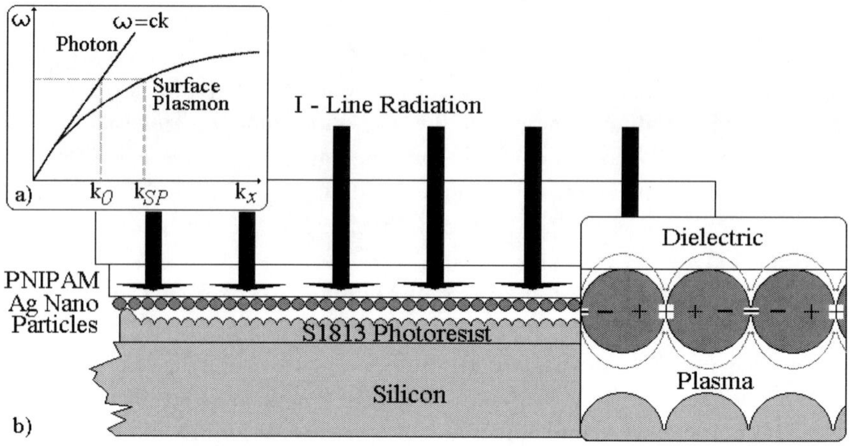

Figure 6. a) Schematic drawing of the surface plasmon dispersion curve. At the same frequency, surface plasmons display a shorter modal wavelength compared to that of free space photons, therefore, allowing sub – wavelength lithography, b) Schematic of the nanolithography process using the Silver mask. Inset: The charges and the electromagnetic field of surface plasmons propagating on the grating surface in the x direction.

For an example structure, a 2D grating coupler (shown in figure 6b) can act as a 2D square array of holes. Light can excite SP waves on the incident side, by selecting a proper periodicity for the hole array and a proper dielectric constant for the medium surrounding the metal substrate, the SP waves can be resonantly coupled through the periodic hole array to the other side of the metal [15]. The SP waves on the exit plane are scattered on the hole array and will be converted back into light waves, if the dielectric constant of the dielectric layers match on both sides of the metal layer. The light wavelength that excite the SP resonance mode (when the incident light is normal to the 2D hole array) is given by [15],

$$\lambda_{(i,j)} = \frac{a}{\sqrt{i^2 + j^2}} \sqrt{\frac{\varepsilon_d \varepsilon_m}{\varepsilon_d + \varepsilon_m}} \qquad (2)$$

where λ is the light wavelength in vacuum, a is the hole array period, i and j are lattice mode indices. The transmission of SP waves through the sub-wavelength plasmonic masks at the I-line wavelength has potential to pattern nanoscale features using conventional near UV light sources. The configuration of the SPR optical lithography is illustrated in Figure 6b. The surface charge displacements depicted in the inset of figure 6b shows the electric field lines hopping between the Ag nanoparticles. However, there are two SP modes and these are illustrated in Figure 7.

The SP modes indicate that the electric field lines for mode 1 at the edges the electromagnetic fields are most concentrated, however, for mode 2 the centers of the both the ridges and the grooves the fields are at maximal. A stronger coupling between the front and back sides of the metal is expected for mode 2. The oscillating current is not dependent of the depth of the grooves, for mode 1 as it is longitudinal (with respect to the wave vector) by hopping over the grooves. It can be seen clearly for mode 2 it is not longitudinal, due to the charge displacement in and out of the grooves and the coupling to the incident light is diminished, when the depth of the grooves is increased, no charge will flow into them.

The transmission of SP waves through the sub-wavelength plasmonic masks at the I - line wavelength has a potential to pattern nanoscale features using conventional near UV light sources. A plasmonic mask designed for the UV wavelength is composed of self assembled silver 30nm diameter nanoparticle hexagonally close packed (hcp) 2D array.

Figure 7. Model of a metal surface when excited by SPR.

2. Fabrication of the Plasmonic Mask

The plasmonic mask was prepared by fabricating a 30 nm diameter hexagonal close packed silver nanoparticle arrays with a period of 50 nm via a self assembly technique. Silver was chosen because in order for the surface plasmon phenomena to occur, it is a pre requisite that a collection of free electrons oscillates at its surface. Hence, a highly conducting metal was chosen. Silver also has the appropriate optical properties at 365nm (I-line), which is the wavelength that is of most interest. To achieve this goal of fabricating the hexagonal close packed 2D array, a hydrophobic surface chemistry was applied to the silver nanoparticles to nucleate them to the surface of the water. Converting the silver nanoparticles from a hydrophilic surface state to a hydrophobic surface state was achieved by using an alkanethiol coating chemistry [18]. Take 0.1ml of the hydrophobic suspension of silver nanoparticles and place them inside a PTFE cylinder using a pipette, fill the PTFE cylinder with deionised water to top. Top up the PTFE cylinder with more deionised water so that the surface of the water is convex. This will bring the silver nanoparticles to the surface, due to the nature of the hydrophilic/hydrophobic interface. Place a Foamvar coated transmission electron microscope (TEM) grid on to the surface of a 2.4cm x 2.4cm borosilicate glass cover slip and cover the whole cover slip (including the TEM grid) with Poly(N-isopropylacrylamide) PNIPAM polymer using a pipette. The verification of the formation of the plasmonic mask was preformed using a TEM. However, in order for the electron beam to penetrate the sample; the sample thickness must be of the order of tens of nanometers, (here approximately 70nm). This is only a guide because the accelerating voltage can be increased to compensate for the increased sample thickness. The PNIPAM polymer thickness was approximately a 100μm, therefore, the PNIPAM polymer was etched to a thickness of approximately 40nm (hence, the silver nanoparticle diameter is 30nm) using an oxygen plasma. The pressure on the vacuum was $1.6x10^{-1}$ mbar, as this was adjusted by flowing Oxygen gas into the vacuum chamber. A current of 80mA was induced to cascade the plasma, the film was etched for a duration of 12min 10 seconds to achieve an approximately 40nm film, this was measured with the TECNAI TEM.

Using the Langumir-Schaefer technique the silver nanoparticles were extracted on to the surface of the PNIPAM polymer via a hydrogen bonding mechanism. The silver nanoparticles / PNIPAM sample was placed in an oven at 56ºC for 24 hours in order to bring the nanoparticles closer together. The TEM grid was cut away (with the silver nanoparticles / PNIPAM on its surface) from the borosilicate glass cover slip using a scalpel and the topography of the plasmonic mask was characterized by using the TEM, operating at 60 KeV. Details of this entire process can be found in reference 18.

3. Detection of Surface Plasmons on Ag thin Film

This plasmonic mask behaves as a perfect asymmetric lens due to the fact that its structure consists of Borosilicate glass, PNIPAM polymer and Ag (which is interfaced to air). The condition of the perfect lens operates, when the permittivity and the permeability are negative [13]. However, at optical frequencies, it is clear that to fabricate a material with both negative ε and μ at the same frequency is quite difficult, due to the lack of magnetic materials. However, it is not necessary to satisfy the condition of $-\mu$, as only - ε of the conditions may be negative because this component of the electromagnetic wave enables the amplification of evanescent waves at both interfaces of the Ag as this leads to the interaction of surface plasmons that is essential to the achievement of the sub-wavelength Photolithography.

The surface plasmon resonances are excited on each Ag interface, this forces the fields of the surface plasmon to interact destructively, therefore this cancels out the incident fields. Hence, zero reflectivity. It is this relation that governs the surface plasmons that is responsible for the perfect lens action. Therefore it is necessary to detect SPR phenomena.

To detect the SPR phenomena in the silver nanoparticles hcp 2D array, the following set up was used to simulate the role of surface plasmons in the plasmonic mask. A Silver (Ag) thin film was deposited on the surface of 2.4cm x 2.4cm borosilicate glass by the sputter coating process. The pressure in the chamber was adjusted by flowing Argon gas into the vacuum chamber, this was $1.6x10^{-1}$mbar and the induced current was 30mA, the film was sputtered coated for a duration of 120 seconds to achieve a 30nm film, whereas by increasing the time the Ag thin film thickness was also increased. Sputtering times of 160s, 200 s and 240s gave Ag thin films thickness of 40nm, 50nm and 60nm respectively. A prism was put in contact with the Borosilicate glass surface via index matching fluid (Edmund Optics). A Helium Neon (HeNe) laser (λ=635nm) was used as a light source. This was pointed at the Ag thin film through the prism / Borosilicate glass. A Silicon photodiode detector was used to detect the reflection that is directed back through the prism at a particular angle as shown in Figure 8.

The sample (Ag thin film and the prism) was rotated from an angle of 0° to 180°. The reflectivity was measured by rotating the detector along with the sample, when the detected radiation intensity decreased the surface plasmon resonance was detected. This was achieved in a dark room.

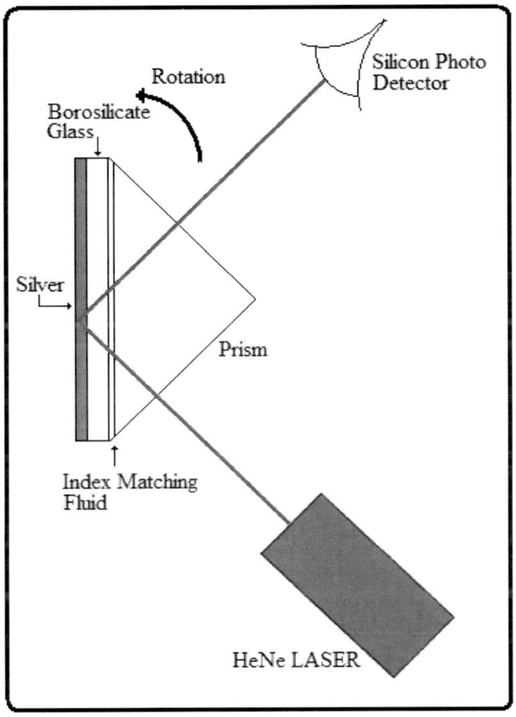

Figure 8. Experimental configuration for the detection of the SPR propagating on a metal thin film.

Figure 8 illustrates the set up of this experiment by having a three layer system; Borosilicate glass (ε_0) which is the dielectric medium (1.536), the silver film (-2.3) [17] (ε_1) of thickness d; and a second dielectric, which in our case was air (1) (ε_2).

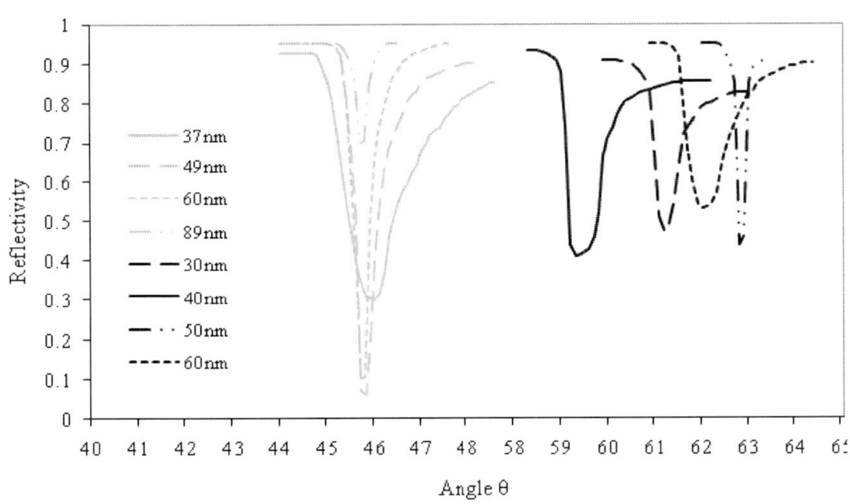

Figure 9. Reflection of Silver films of different thickness. Detection of Surface Plasmon modes, the data represented in black were performed at the 635nm wavelengths in comparison with the data of [17] at 546nm shown in grey.

Figure 9 shows the effect of variation of thickness at a constant wavelength (λ = 635nm) on the position of the attenuated total reflection (ATR) minimum. The radiation wave is partially reflected at the borosilicate glass / silver film interface, having passed the prism (ε_0). It passes through the silver film of thickness *d* as an exponentially decaying wave as depicted in figure 10 [19]. At the Silver film / air interface it induces excitations which radiate light back into the silver film.

If the thickness *d* increases, the backscattered field disappears due to the decreasing transmission. Therefore the reflectivity interaction becomes more dominant. However, for a decreasing thickness *d* the back scattered field increases, where the optical transmission increases and the reflectivity has a minimal interaction. The incoming wave is in antiphase with the back scattered field, therefore, the reflection is reduced as the two interfere destructively. The antiphase radiation field, captured in the Ag film, is converted into heat.

For the surface plasmon resonance case, the decrease in radiation reflected, amounts to absorption, when there are no reflections. This is due to the radiation emitted from the SP's at the sliver film / air interface back into the borosilicate glass in the direction of the reflected beam.

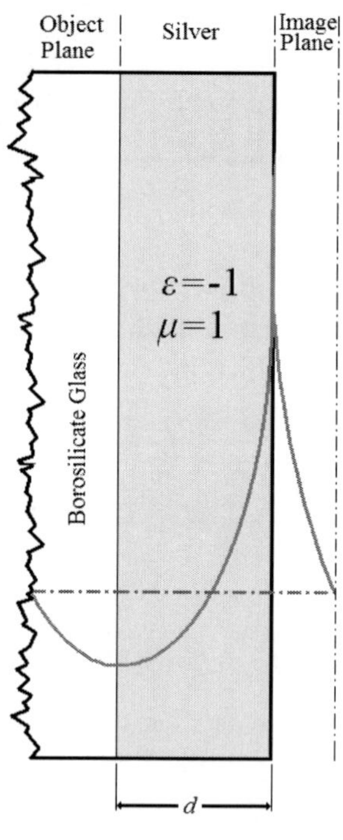

Figure 10. The perfect asymmetric lens system consisting of a thin film of silver.

If the silver film thickness *d* is decreased further this leads to increasing optical transmission and this has a direct effect on the reflectivity, where it's increasing role become more dominant. A quantitative explanation of the minimum of the reflected intensity was proved by Raether [17]. It was calculated from the surface plasmon dispersion equation (see Figure 6a) that the reflections would come to a minimum at an angle of 62°. The surface plasmon dispersion equation can be expressed as [17].

$$k_{sp} = \frac{2\pi}{\lambda}\sqrt{\frac{\varepsilon_m \varepsilon_d}{\varepsilon_m + \varepsilon_d}} \tag{3}$$

where λ is the wavelength of the incident radiation, which in this case was 635nm. The dielectric constant of a metal and a surrounding dielectric material are represented by ε_m and ε_d, respectivitely. Due to experimental error, not all the data has dipped to an angle of 62°. This has probably occurred due to the fact that the incident radiation has not entered the prism at exactly the same angle and excited the SPR at the same place as the previously measurement, therefore, the conditions have altered slightly when the sample was rotated. However, surface plasmons have been detected near the desired angle.

3.1. Optical Properties of the Hexagonally Closely Packed Ag Nanoparticle Array

For a photolithography optical mask there must be a difference in contrast between the Borosilicate glass and the (30nm) Silver nanoparticles, where the glass is the light part and the Silver nanoparticle is the dark part of the optical mask. This difference was determined using a Jenway spectrophotometer and the data is displayed in Figure 11.

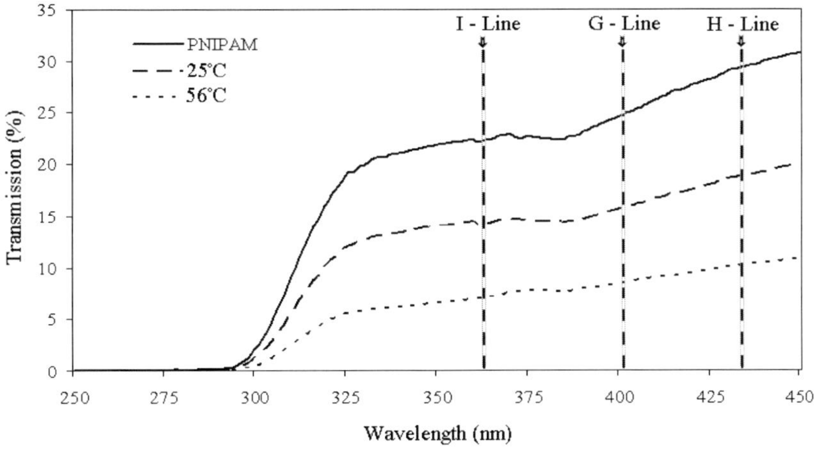

Figure 11. Transmission spectrum of the 30nm spherical Silver nanoparticle with a spacing period of 20nm forming a hcp 2D array.

The transmittance spectra showed that the borosilicate glass without silver nanoparticles was approximately 90% transparent in the UV – radiation wavelengths [18]. Therefore, this was a perfect substrate to produce an opaque plasmonic photo mask. It is observed that the PNIPAM polymer attenuated approximately 75% of the UV radiation at the I–line wavelength this is due to the DEAP photoinitiator. The DEAP cures at lower wavelengths but is still effective at the I–line wavelength.

As the distance between the silver nanoparticles decreases the optical transmission decreased, this is due to the SPR which the optical properties of silver nanoparticles are determined by a collective oscillation of free electrons in the nanoparticles which is mostly described by the term plasma – resonance absorption. The plasma – resonance absorption depends on the nanoparticle material and on nanoparticle structure, size and shape, in this case the Ag nanoparticles were 30nm diameter and spherical. The observed transmission, spectra strongly depends on the wavelength and the period of the hcp 2D array, which confirms the role of SPs.

The PNIPAM was heated at a temperature of 56°C for 24 hours the PNIPAM shrank in all directions as the Silver nanoparticles on the surface were brought closer together as shown in Figure 12a.

Figure 12. a) Silver nanoparticle hcp 2D array positioned by self assembly, b) developed S1813 photoresist exposed at I – line wavelength produced by surface plasmons.

From the data shown, it is evident that the silver mask is more opaque as a function of temperature. It is clear that the silver nanoparticle hcp 2D array does not absorb all of the UV radiation, 7.13% at the I – line wavelength, 8.78% at the G – line wavelength and 10.32% at the H – line wavelength. However, it is evident that from Figure 12b, the absorption mechanism (SPR at the surface of each silver nanoparticle) is present as the features remain. This will affect the resultant thickness of the exposed photoresist, as the transmission of the UV radiation increases then the resultant photoresist thickness will decrease. Therefore, G and H – line wavelengths will result with a thinner photoresist thickness.

4. Direct Nano Patterning of a Two Dimensional Structure

In the Photo–Nanolithography process, Shipley S1813 photoresist was spun directly on to a Foamvar coated TEM grid at 6000 rpm for the duration of 45 s. This resulted in a uniform thickness of 1.12μm. In conventional photolithography, the edge beads are not required as the optical mask is held directly in contact with the photoresist. However, as shown in Figure 13, the SPR mask was suspended with the silver nanoparticle side down on the edge beads, which is advantageous as this reduces the number of steps in the lithography process. An 8 s exposure was preformed, using a filtered mercury vapour lamp with a radiation peak at 365 nm. The exposure area was dependent upon the SPR mask, where the silver nanoparticles were ordered into a hexagonal close packed array.

The S1813 photoresist was developed after UV exposure using Microposit 2401 developer and was then rinsed in deionized water. The development time was 15 s. The topography of the exposed features was characterized by using the TEM, operating at 60 kV.

In order for the Plasmonic mask to function as an asymmetric perfect lens to achieve sub – wavelength photolithography, Figure 12a shows a highly ordered high density silver nanoparticle hexagonally closely packed 2D grating that was fabricated. This shows suitable periodic structure that can support the phenomena of SPR.

The TEM image shows feature sizes of 30 nm with a period of 50 nm is uniform. The TEM image shown here does not look crisp as this is down to experimental error as the TEM was not focused correctly. However, it can be seen clearly that the resolution of silver nanoparticle SPR mask is approximately 12 times smaller than the I-line wavelength (λ /12).

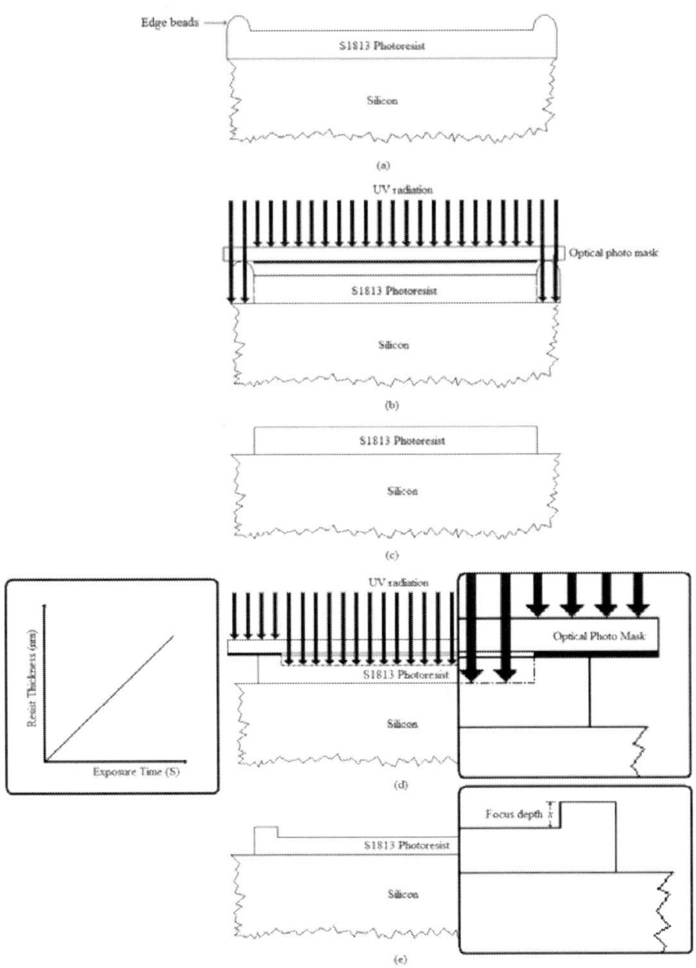

Figure 13. Schematic of lithographic process to obtain the correct edge bead thickness to focus the plasmonic mask. a) edge beads occurring when spun at a particular speed, b and c) UV exposure to get rid of the edge beads and develop, d) UV exposure as a function of time, inset, close up showing the distance of the UV exposure penetrating the photoresist as a function of time, e) developed photoresist, inset, resultant thickness to focus the plasmonic mask.

The results shown in the TEM micrograph of Figure 12b, the developed photoresist features are on average 34 nm in diameter with a smaller period of 46 nm than the features produced by the plasmonic mask, therefore, the S1813 photoresist features are 11 times smaller than the wavelength of the I-line photons (λ/11). This pattern period is far beyond the diffraction limit of far-field lithography [18, 20].

From this it is evident that the evanescent electromagnetic field of the surface plasmon at the silver nanoparticle / borosilicate glass interface propagates across the silver nanoparticle and is superimposed on the field of the surface plasmon oscillating at the same frequency at the silver nanoparticle / air interface. The field is leaked into the air gap as depicted in the inset of Figure 13b [18, 21]. The transmitted leaked field (in this case the radiation λ = 365nm) interacts with the S1813 photoresist and produces the features shown in Figure 12b.

Due to the quantity of leaked UV radiation the developed S1813 photoresist thickness was approximately 33 nm (G and H-Line would result in thinner photoresist, as the difference between light and dark areas (Ag nanoparticles) has become less, see Figure 12).

The optimal exposure time is only 8 s, this is comparable to a typical exposure dose used in conventional lithography with larger features, which implies a strong near-field transmission enhancement due to SPs.

The S1813 photoresist features are 34nm and not 30nm because the plasmonic mask lens was out of focus as the thickness of the edge beads, were either too high or too low. This can be rectified by the edge beads either being higher or lower either by spinning the photoresist at a faster / slower spin speed or by optimizing the timing of the UV exposure so that the S1813 photoresist has a resultant lip so that the plasmonic mask lens can sit directly upon on it as illustrated in Figure 13.

CONCLUSION

A novel plasmonic photolithography using a periodic hexagonally closely packed silver nanoparticle array mask has been demonstrated. To confirm the existence of surface plasmons propagating at the silver surface, it was found that the SPR was detected at an angle of 62° of the silver film with thicknesses ranging from 30 – 60nm. Far – field transmission spectra of the hexagonal structures shows strongly enhanced transmission in the UV range, which opens the possibly of lithography using the apertures smaller than the exposing wavelengths. At temperatures of 25°C the optical transparency of the silver nanoparticles was 14% at I-line wavelengths, whereas when the temperature was elevated to 56°C the optical transparency was increased to 7.13%. This was due to the fact that as the temperature increases, the PNIPAM polymer shrinks bringing the silver nanoparticles on its surface closer together to form a periodic 2D grating, therefore, increasing the plasmonic mask opaqueness.

The nanolithography technique utilizes near UV radiation to excite surface plasmons on a metal grating in order to enhance the transmission through sub wavelength periodic apertures with effectively shorter wavelengths compared to the excitation light wavelength. The plasmonic mask consisted of the ordered silver nanoparticle hexagonal arrays with feature sizes of 30 nm with a period of 50 nm, where the resolution of silver nanoparticle plasmonic mask was approximately 12 times smaller than the I-line wavelength (λ/12). Nanolithography was performed with an optimal I-line exposure time of 8 seconds.

The result of this exposure showed the transfer of the 2D hexagonal array pattern into the photoresist. Here the inter-particle pitch was 46nm and the features of the developed S1813 photoresist had a diameter of 34nm. This was found to be 11 times smaller than the wavelength (λ/11).

As well as arrays, so called nanowires were experimented with. The nanowire widths varied between approximately 70nm – 10μm and the lengths were tens of microns. These nano structures could be controlled by decreasing the pressure applied to the Ag nanoparticles. When aligning the nanowires, they had a tendency to distort and bend as they were compressed together and even in some instances the nanowires fractured into small chains. Whilst the suitability of this technique in isolation may not be compatible with high throughput lithography, combining it with e-beam lithography for very high resolution features shows some promise.

REFERENCES

[1] G. E. Moore, 'Cramming more components onto intergrated circuits', *Electronics*, 38 (80 (1965).

[2] R. Mahajan, R. Nair, V. Wakharkar, J. Swan, G. Vandentop, 'Emerging directions for packaging technologies', *Intel Technology Journal*, 6 (2), pp. 62 – 75, (2002).

[3] H. Bakoglu, 'Circuit, interconnection and packaging for VLSI', Addison – Wesley publication Company, Reading MA, (1990).

[4] M. D. Levenson, D. S. Goodman, S. Lindsey, P. W. Bayer, H. A. E. Santini, 'The phase – shifting mask II: imaging simulations and sub micrometer resist exposure', *IEEE. Trans. Electron. Dev*, vol ED – 30 (6) pp. 753 – 763 (1984).

[5] L. R. Harriot, 'Limits of Lithography', *Proc. IEEE*, Vol 89, issue 3 pp 366 – 374 (2001).

[6] F. M. Schellenberg, H. Zhang, J. Morrow, 'Sub wavelength lithography using OPC', SEMATECH J111 project OPC validation, *Proc SPIE,* 3334, pp. 892 – 911 (1998).

[7] Erdmann, W. Henke, S. Robertson, E. Richter, B. Tollkuhn, W. Hoppe, 'Comparison of simulation approaches for chemically amplified resist', *Proc SPIE*, 4404, 99 (2001).

[8] S. Rizvi, 'Handbook of photomask manufacturing technology', CRC Taylor and Francis group pp 139 (2005).

[9] H. Schmid, H. Biebuyck, B. Michel, O. J. F. Martin,'Light-coupling masks for lenses, sub-wavelength optical lithography', *Appl. Phys. Lett.* 72, 2379 (1998).

[10] J. G. Goodberlet, 'Patterning 100nm features using deep-ultraviolet contact photolithography', *Appl. Phys. Lett.* 76, 667 (2000).

[11] J. G. Goodberlet, H. Kavak, 'Patterning sub-50nm features with near-field embedded-amplitude masks', *Appl. Phys. Lett.* 81, 1315 (2002).

[12] O. J. Martin, N. B. Piller, H. Schmid, H. Biebuyck, B. Michel, 'Energy flow in light-coupling masks for lensless optical lithography', *Opt. Express* 3, 280 (1998).

[13] M. M. Alkaisi, R. J. Blaikie, S. J. McNab, R. Cheng, D. R. S. Cumming, 'Sub-diffraction-limited patterning using evanescent near-field optical lithography', *Appl. Phys. Lett.* 75, 3560 (1999).

[14] T. W. Ebbesen, H. J. Lezec, H. F. Ghaemi, T. Thio, P. A. Wolff, 'Extraordinary optical transmission through sub-wavelength hole arrays', *Nature* (London) (1998), 391, 667 – 669.

[15] T. Thio, H. F. Ghaemi, H. J. Lezec, P. A. Wolff, T. W. Ebbesen, 'Surface-plasmon-enhanced transmission through hole arrays in Cr films', *JOSA B* (1999), 16, 1743 – 1748.

[16] L. Salomon, F. Grillot, A. V. Zayats, F. deFornel, 'Near-field distribution of optical transmission of periodic subwavelength holes in a metal film', *Phys. Rev. L*ett. (2001), 86, 1110 – 1113.

[17] H. Raether. Surface Plasmons on Smooth and Rough Surfaces and on Gratings, Springer, Berlin, (1988), pp 25 – 30.

[18] S. Lewis, R. Wheeler – Jones, V. Haynes, R. M. Perks, **'**High density self assembled nanoparticle film with temperature-controllable interparticle spacing for deep sub-wavelength nanolithography using localized surface plasmon modes on planar silver nanoparticle tunable grating', *Micro. Eng*., 85 (2) pp. 486 – 491 (2008).

[19] J. B. Pendry, 'Negative refraction makes a perfect lens', *Phys. Rev. Lett.* 85 3966, (2000).

[20] W. Srituravanich, S. Durant, H. Lee, C. Sun, X. Zhang, 'Deep subwavelength nanolithography using localized surface plasmon modes on planar silver mask', *J. Vac. Sci. Technol*. B 23 (6) (2005).

[21] J. Seidel, F. I. Baida, L. Bischoff, B. Guizal, S. Grafström, D. Van. Labeke, L. M. Eng, 'Coupling between surface plasmon modes on metal films', *Phys. Rev. B* 69, 121405 (2004).

[22] S. W. Chung, G. Markovich, J. R. Heath, 'Fabrication and alignment of wires in two dimensions', *J. Phys. Chem. B* (1998), 102, 6685.

In: Plasmons: Structure, Properties and Applications
Editors: Anton E. Turunen and Johannes O. Niemi
ISBN: 978-1-62100-060-0

Chapter 5

Applications of Gold Nanorods Based on Localized Surface Plasmon Resonance

Yinan Zhang and Yu Chen
Photophysics Group, Centre for Molecular Nanometrology,
Department of Physics, SUPA, University of Strathclyde,
John Anderson Building, Glasgow, UK

Abstract

Metal nanoparticles possess localized surface plasmon resonances (LSPR), which leads to unique effects, such as multiphoton-induced luminescence, radiative quenching or enhancement and spaser. In particular, gold nanorods (GNRs) are potential candidates for examining all these LSPR effects. GNRs have longtudinal surface plasmon band in the near infra-red (NIR) region, where the absorption of water and biological tissue is minimized, making them useful in biological research.Moreover, this longitudinal plasmon band can be tuned by changing the size, shape of nanorods and the dielectric environment. In this chapter, we review our work on surface plasmon enhanced two-photo luminescence (TPL) and energy transfer effect. The application of GNRs in biomedical imaging and analytical studies has been examined, together with a survey on localized surface plasmon related fluorescence enhancement and recent development on spaser.

1. Introduction

The surface plasmon oscillation is the coherent excitation of free electrons within the conduction band, leading to an in-phase oscillation. In the case of surface plasmon resonance (SPR), plasmons propagate in the directions along the metal-dielectric interface, for distances on the order of tens to hundreds of microns, and decay in the perpendicular direction with 1/e decay lengths on the order of 200nm.

There have been varieties of applications related to SPR, including surface-enhanced spectroscopy,[1-7] biological and chemical sensing,[8-16] and lithographic fabrication,[17-20] etc al. While with progress in fabrication and manipulation of metallic materials, especially on nanoscale, localized surface plasmon resonance (LSPR) has attracted much attention.

For the case of localized surface plasmons, incidence electromagnetic (EM) wave interacts with particles much smaller than the incident wavelength, leading to a charge density oscillation confined to metallic nano structures. Excitation of LSPR by EM field, such as light, where resonance occurs, provide strong light scattering in the appearance of intense surface plasmon absorption bands and enhancement of the local EM field.[20]

Many applications can be achieved by introducing LSPR, especially in spectroscopy research, such as near-field scanning optical microscopy,[21-23] surface-enhanced Raman spectroscopy (SERS),[24-26] multiphoton-induced luminescence,[27-29] radiative quenching or enhancement[30-34] and spaser.[35-37] Gold nanorods (GNRs) are potential candidate for examining all these SPR effects.

The surface plasmon bands of GNRs split into two parts: transverse and longtudinal band. The longitudinal one resonates in the near infra-red (NIR) region, where the absorption of water and biological tissue is minimized, making them useful in biological research. Moreoever, this longitudinal plasmon band can be tuned by changing the size, shape of nanorods and the dielectric environment.[38]

In this review we will examine LSPR effect on luminescence process and their applications in biomedical imaging and analytical studies, as well as recent development on spaser.

2. Localised Surface Plasmon Related Luminescence |of GNRs

2.1. Localised Surface Plasmon of Gold Nanorods

The electric field of incidence light induces a polarization of the free conduction electrons. As the ionic core is much heavier than the free electron, net charge will occur at the nanoparticle boundaries acting as a restoring force. Then a dipolar oscillation of electrons is created.

For spherical particles, the surface plasmon induced absorption can be described by Mie theory [38]. The expressions are displayed below.

$$\sigma_{ext} = \frac{2\pi}{|k|^2} \sum_{L=1}^{\infty} (2L+1) Re(a_L + b_L) \quad (1)$$

$$\sigma_{ext} = \frac{2\pi}{|k|^2} \sum_{L=1}^{\infty} (2L+1)(|a_L|^2 + |b_L|^2) \quad (2)$$

$$\sigma_{abs} = \sigma_{ext} - \sigma_{sca} \quad (3)$$

In the equations above, σ_{ext}, σ_{sca}, *and* σ_{abs} represents extinction, scattering and absorption cross-section separately. a_L and b_L are functions of r, ε_m and $\varepsilon(\omega)=\varepsilon_1(\omega)+i\varepsilon_2(\omega)$, where r is the radius of the nanopraticle, ε_m and $\varepsilon(\omega)$ are the dielectric functions of the surrounding medium and the material itself respectively. For nanoparticles much smaller than the wavelength of incidence light, only the dipole oscillation contributes significantly. The Mie theory can be reduced to the following equation:

$$\sigma_{ext}(\omega) = 9\frac{\omega}{c}\varepsilon_m^{3/2}V\frac{\varepsilon_m(\omega)}{[\varepsilon_1+2\varepsilon_m]^2+\varepsilon_m(\omega)^2} \quad (4)$$

with V being the particle volume, ω the angular frequency of the incidence light and c the speed of light. The resonance condition is fulfilled when $\varepsilon_1(\omega)=-2\varepsilon_m$ only if ε_2 is weakly dependent on ω. The above equation can be used to explain the absorption spectra of small gold nanoparticles. For large particles, the dipole approximation is not valid and higher-order modes of oscillation have to be taken into account, which will cause the absorption spectrum peaking at lower energies.

For gold nanorods, the absorption spectrum splits into two bands with one at around 500nm corresponding to the oscillation of the electrons perpendicular to the long axis and the other at lower energy representing the oscillation of the free electrons along the long rod axis. These two modes are referred to as the transverse and longitudinal plasmon surface absorption. The scheme of surface plasmon induced absorption band is shown in Figure1.

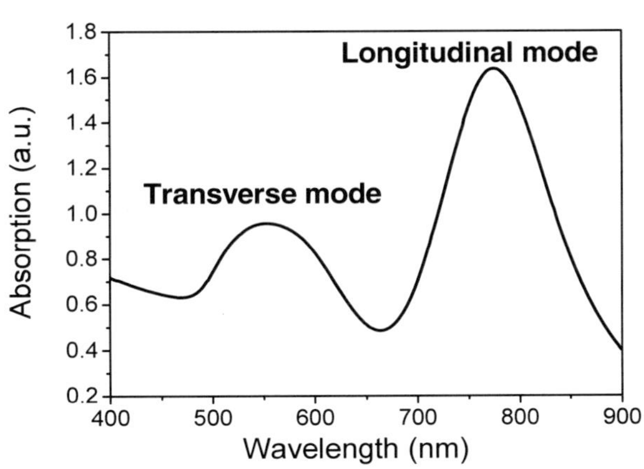

Figure 1. A typical absorption spectrum of gold nanorods with surface plasmon modes as labelled.

The absorption spectrum can be described by an extension of the Mie theory in a quantitative manner.

The extinction cross-section is

$$\sigma_{ext}(\omega) = \frac{\omega}{3c}\varepsilon_m^{3/2}V\frac{(\frac{1}{P_j^2})\varepsilon_2}{\{\varepsilon_1(\omega)+[\frac{1-P_j}{P_j}]\varepsilon_m\}+\varepsilon_1^{2}} \quad (5)$$

where P_j are the depolarization factors along the three axes A, B and C of the nanorod with A>B=C, defined as

$$P_A = \frac{1-e^2}{e^2}\left[\frac{1}{2e}\ln\left(\frac{1+e}{1-e}\right)-1\right] \tag{6}$$

$$P_B = P_C = \frac{1-P_A}{2} \tag{7}$$

$$e = \left[1-\left(\frac{B}{A}\right)^2\right]^{1/2} \tag{8}$$

R is the aspect ratio describing the value of the long axis divided by the short axis of a rod-shaped particle. In analogy with the absorption of spheres, the plasmon resonance condition is:

$$\varepsilon_1(\omega)+\left(\frac{1-P_j}{P_j}\right)\varepsilon_m = 0 \tag{9}$$

Therefore, nanorods with given aspect ratio have two absorption bands, one corresponding two R_B=R_C and the other to R_A.

The localized surface plasmon structure of gold nanorods is strongly dependent on the size, shape and surrounding environment of the nanoparticle. Figure 2(a) shows extinction spectra of nanorods with different aspect ratios, normalized by the transverse surface plasmon band at around 550nm. All gold nanorods were synthesized by a seeded-growth method.[39] The longitudinal surface plasmon band shifts from around 700 nm to 1000 nm as the aspect ratio increases from 2 to around 5. Figure 2 (b) shows a typical SEM image from gold nanorods with longitudinal mode centred close to 700nm. The ratio of the longitudinal to transverse band also increases with the aspect ratio, indicating the former mode dominates in 'longer' nanorods.

2.2. Surface Plasmon Enhanced Two-Photon Luminescence (TPL)

The photon induced luminescence from gold was first reported by Mooradian.[40] The photoemission is thought to arise from the direct transitions between electrons in the conduction band below the Fermi level and holes in the d bands. But the single photon luminescence efficiency is very low, around 10^{-10}. Luminescence of noble metal materials did not attract much attention until Boyd reported multiphoton-induced luminescence on roughened noble metal surfaces, which can induce luminescence several orders of magnitudes stronger than single photon emission.[41] This amplification of luminescence is believed due to a resonant coupling with localized surface Plasmon,[42] which has broad applications in fluorescence research. In addition to enhanced luminescence intensity, multi-photon excitation provides unique advantages compared to single photon excitation. Multi-photon process has non-linear dependence on incidence power, which will provide better on-axis spatial resolution.Longer excitation wavelength also provides deeper penetration depth in biological tissue. For GNRs luminescence research, two-photon excitation at near-infrared band are commonly applied, the energy window through which damage to biological system and tissue auto-fluorescence can be minimized.

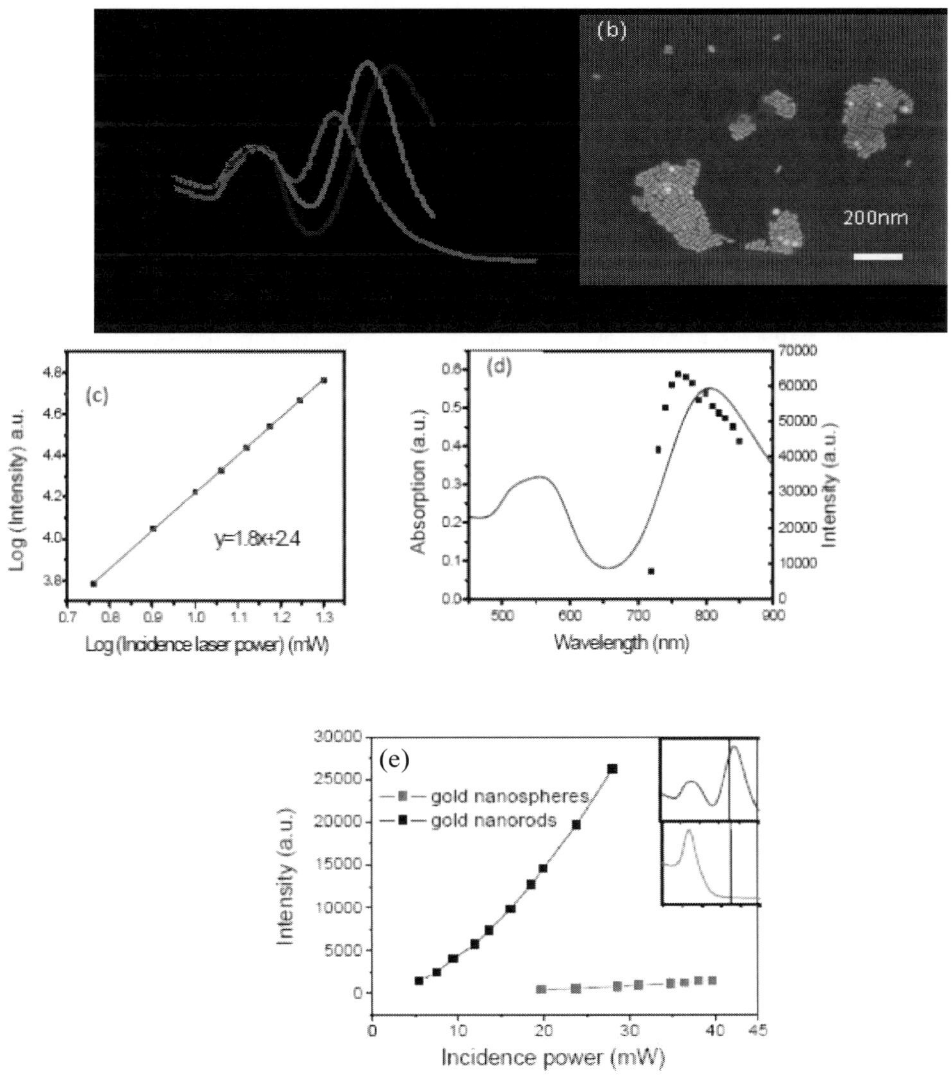

Figure 2. (a) Extinction spectrum of nanorod samples normalized at transverse band; (b) SEM image of GNRs; (c) Fluorescence intensity dependence on incidence laser power (in log scale); (d) TPL excitation spectrum (solid squares) compared with absorption of gold nanorods (solid line); (e) comparison of fluorescence intensity from gold nanorods (black squares) and nanospheres (red squares) excited at 750nm, inset: normalized absorption spectrum of nanorods (black line) and nanospheres (red line), the vertical black line indicates the excitation wavelength.

All gold nanorods solution in Figure 2(a) show strong fluorescence under NIR excitation. The nonlinear nature of the TPL signal was confirmed by measuring the dependence of the luminescence intensity as a function of excitation power on all samples. A typical result is shown in Figure 2(c), where data was collected from nanorods with a longitudinal band at around 800nm (blue curve in Figure 2(a)) under 800nm excitation and the incidence power increases from 5 to 20mW. The dependence of fluorescence intensity on incidence laser power was found close to quadratic, suggesting a non-linear, two-photon excitation process.

A. Surface Plasmon Effect

TPL from GNRs is found governed by the longitudinal plasmon mode. Figure 2(d) displays the TPL intensity (black squares) from nanorods excited under constant laser power but different wavelength varying from 720nm to 850nm, together with the extinction spectrum (solid line). The excitation spectrum overlaps well with the longitudinal plasmon band, indicating that the TPL intensity is governed by the local field enhancement from the plasmon resonance. A slightly blue shift between the excitation profile and the absorption spectrum was observable, possibly due to the slightly descent in detector efficiency approaching to IR band. The governance behaviour of TPL by LSP was also observed by Wang et al. They measured the relative TPL efficiency of several types of GNRs of which aspect ratios ranging from 1.3 to 4.3 (meaning longitudinal modes from 54onm to 820nm) at 815nm excitation. The result shows that GNRs with longitudinal mode most close to the excitation wavelength (in their experiment, the one with longitudinal band at 820nm) have the greatest TPL efficiency. The enhanced absorption by surface plasmons is thought to be the main contribution to the enhanced TPL efficiency, which increases the pumping rate of electron-hole pairs.[43]

Furthermore, the TPL from gold nanorods and nanospheres were compared in Figure 2(e). The difference in absorption spectrum can be clearly identified in the inset of Figure 2(e), where nanorods have a significant longitudinal band in the NIR region. With excitation at 800nm (marked by the solid line in the absorption spectrum), under same excitation power, TPL intensity from nanorods is about 30 times stronger than that from nanospheres, which suggests that TPL from gold nanorods is mainly governed by the longitudinal surface plasmon band.

B. Polarization Dependence

The intensity of TPL from gold nanorods is dependent on polarization of the incidence beam. TPL enhancement is maximized when the incident field is parallel with the longitudinal dipolar plasmon resonance along the long axis of the nanorod. This polarization dependence property makes it possible to identify single nanoparticles from particle clusters, as the latter provides similar TPL intensity under different incidence excitation orientation while single particle does not. The polarization dependence behaviour on incidence excitation is shown in Figure 3, in which GNRs were excited under perpendicular incidence excitation polarizations. Luminescence from single nanorod can be identified where luminescence vanishes when the incidence polarization is changed. Spots showing average luminescence intensity under both excitation orientations are considered to be nanorods clusters. Quantitative measurement by Wang et al. reveals that the TPL intensity I follows the relationship $I=I_{in}cos^4\theta$, where I_{in} is the incidence laser power and θ represents the angle between the incidence beam polarization and the long axis of gold nanorods.[29] The TPL from a single nanorod has the same polarization dependence as the two-photon fluorescence from a single fluorescent molecule, in accord with the dipolar nature of the LSP of gold nanorods.[29, 44] However, the TPL emission from gold nanorods is depolarized, different from plasmon-resonant scattering processes, which retain the polarization of the incident beam. Furthermore, the TPL emission spectrum appears to be very broad and is likely to be produced directly by electron–hole recombination rather than through intermediate particle plasmons in the single-photon luminescence of gold nanoparticles.

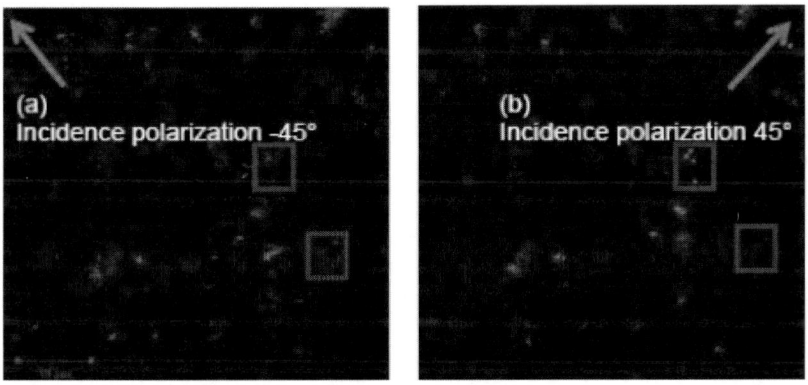

Figure 3. TPL images under perpendicular excitation polarizations.

Therefore, the plasmon-enhanced TPL from nanorods is essentially due to a greater two-photon absorption cross section.[29, 44] On the otherhand, for larger nanaorods at sub-micro meter scale, near-field study by Imura et al. reveals a $cos^2\theta$ dependence of TPL intensity on incidence polariation. This relationship was explained by sequential one photon excitation processes: an electron in the sp conduction band is excited by the first photon from below the Fermi surface to above it via an intraband transition, resulting in a charge-separated state within the conduction band. A second photon excites an electron from the d band to the sp band, transferring the hole to the d band. Eventually, an electron-hole recombination emits luminescence. The first transition is maximized along the longitudinal surface plasmon orientation but the polarization information is rapidly lost after excitation, whereas the second transition is polarization insensitive. Therefore, the photoluminescence polarization follows a $cos^2\theta$ relationship, and does not depend on a plasmon enhanced absorption as is the case of the smaller NRs. These observations are not in conflict because the submicron NRs used in the near-field study do not have the appropriate aspect ratio to support a dipolar plasmon resonance at NIR frequencies. [44-47]

C. Fluorescence Lifetime

Fluorescence lifetime is the average time a fluorophore molecule spends in the excited state prior to return to the ground state. It is a critical parameter describing a fluorescence process, as it determines the time available for the fluorophore to interact with or diffuse in its environment. Time-resolved research on fluorescence decay time can provide information on fluorescence dynamics, surrounding environment and interactions of fluorophores with other particles. The lifetime decay curve of TPL from GNRs was measured with GNRs solution under 800nm excitation using home-built time-correlated single photon counting (TSCPC) system. The lifetime is found to be around 100ps but is much longer than those achieved in pump-probe and upconversion research, where the lifetime of excited electrons in gold nanorods is found to be less than several picoseconds.[48, 49] This is probably due to the limit of time resolution caused by the instrumental response function of the TCSPC system, which is also around 100 ps. Our measurements can give the upper limit of the decay time of TPL process. This result is confirmed by FLIM measurement on GNRs samples disposed on glass slides.[50]

2.3. Applications in Biomedical Imaging

Besides their decent TPL intensity, GNPs have several advantages for biological imaging: they are photo-stable, less or even non-toxic to biological system and can be conjugated to various types of bio-molecules.[44, 51-54] Therefore, surface plasmon enhanced TPL from gold nanorods have been applied to biological imaging research in a variety of aspects, both in vitro and in vivo. For example, TPL of GNRs was used to trace the delivery of floate-conjugated nanorods (F-NRs) to KB cells (a tumor cell line derived from human oral epithelium). Microscopy experiments show a high signal-to-background ratio of the TPL contrast generated by internalized NRs.[44, 55] Also there has been research on nonspecific cellular uptake of NRs with different surface coatings based on TPL microscopy, which is a critical issue to clinical applications.[44, 56] One example of microscopy study on uptake of gold nanorods by MDCK cells is presented in Figure 4.[49] The black white TPL image of reference gold nanorods samples dispersed on a glass slide is shown in Figure 4(a).

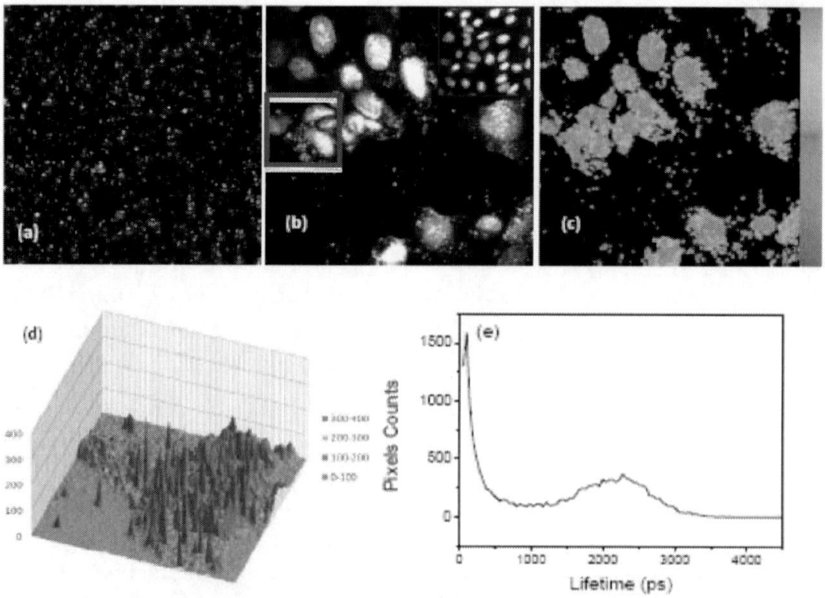

Figure 4. (a) Intensity image of TPL from GNRs disposed on glass slide; (b) TPL intensity and (c) lifetime images of gold nanorods internalized by MDCK cells; inset - intensity image of reference MDCK cells; (d) 3D intensity distribution of the region marked by red rectangle in Figure3 (b) (arbitrary scale); (e) Lifetime histogram in (c).

Bright spots are due to TPL from nanorods, while bigger ones correspond to nanorods clusters. Strong TPL from separated nanorods can be well identified. Figure 4(b) presents a TPL image from MDCK cells incubated with GNRs. The elliptical shapes correspond to nuclei stained by 4'-6-Diamidino-2-phenylindole (DAPI) (Emission maximizes at around 460nm and does not shift much under two photon excitation), while bright spots are due to TPL from gold nanorods. Compared with cell reference sample (inset in Figure 4(b)), it is clear that there was nanorods taken up by MDCK cells, and nanorods showed a stable and strong luminescence in cell environment. Figure 4(d) shows the spatial intensity distribution of the region marked by red rectangle in Figure 4(b), where emission from gold nanorods, represented by sharp peaks, has intensity several times stronger than that from DAPI.

Moreover, beyond traditional confocal microscopy imaging method based on TPL intensity from gold nanorods, lifetime measurement is even more powerful. Different from traditional microscopy, which relies on the contrast that can be achieved from absorption, polarization, phase or fluorescence intensity, fluorescence lifetime imaging microscopy (FLIM) provides contrast according to the fluorescence decay time. The fluorescence decay time depends not only on the intrinsic characteristics of the fluorophore, but also upon the local environment, such as the local viscosity, pH, or refractive index, as well as interactions with other molecules, such as collision or energy transfer. Thus, as well as being able to distinguish spectrally overlapped fluorophores, imaging of the fluorescence lifetime can be used to probe the surroundings of a fluorophore.[50] Figure 4(c) is such a FLIM image taken from the same area as Figure 4(b), where different coded colors represent different lifetime scale. The fluorescence lifetime of each pixel is obtained by applying a single or multi-exponential fit to the intensity image; in this case a single exponential model was used. The fluorescence lifetime of DAPI is found to be longer than 2 ns, and fitted well to a single exponential model. This value is in accordant with published data on FLIM analysis of DAPI. [57] While the TPL decay time of gold nanorods is found to be shorter than 100 ps, and this result is confirmed by a FLIM study of pure gold nanorods on a glass slide as well as time-resolved fluorescence measurement of gold nanorods in a cuvette using the same excitation. The lifetime distribution histogram, Figure 4(e), clearly shows the apparent gap between lifetimes of TPL from GNRs and DAPI. This characteristic short lifetime of TPL from gold nanorods thus can be useful in distinguishing gold nanorods from extrinsic and endogenous fluorophores in multi-labelled biological systems. [50,58]

TPL imaging with gold NRs provides high spatial resolution and penetration depth to monitor biological behaviors in vivo. Wang et al. provided the first example of in vivo TPL imaging by monitoring GNRs flowing through the blood vessels in a mouse earlobe. Picomolar concentrations of CTAB-GNRs were introduced by tail vein injection, and then presumably diluted in the blood pool to femtomolar levels prior to real-time TPL imaging. The TPL signals were approximately three times higher than the background autofluorescence, and the uniform intensities of the emissions in the single-frame image suggest that most of these were generated by single GNRs. No TPL signals were detected 30 min after injection, indicating the clearance of GNRs from circulation. [29, 44]

3. Energy Transfer

Energy transfer is one of the fundamental processes governing light-harvesting in biological systems and energy conversion in electronic devices such as organic solar cells or light-emitting diodes, and it can be applied in various aspects of biological research, such as bio-sensing and imaging. There are two basic types of energy transfer process: Förster resonance energy transfer (FRET) and surface energy transfer (SET). FRET occurs between donors and acceptors in close proximity, is referred to as the 'spectroscopic ruler' as the transfer rate critically depends on the sixth power of the separation between the donor and acceptor. FRET theory assumes a (point)-dipole-dipole interaction between a donor and an acceptor. The effective range of FRET is reported to be typically less than 80Å, as limited by the nature of the dipole-dipole interaction.[59] While surface energy transfer from a dye to metal surface, described by the SET model, has a fourth power distance dependence and therefore an extended effective range.[60] To this end, gold nanoparticles have been applied in energy transfer research. For instance, T. Sen et al. have demonstrated energy transfer phenomenon on a simple mixed solution of Rohdamine 6G and gold nanospheres.[30] A 40%-85% decrease of fluorescence intensity from Rohdamine 6G together with a shortening fluorescence lifetime of the dye have been detected. The results were explained by a surface energy transfer (SET) model, in which the transfer efficiency is dependent on d^4, where d is the separation between the dye and gold nanoparticles. Li et al studied the energy transfer effect based on GNRs with different LSPR peaks conjugated with quantum dots (QDs) to examine the impact of the LSPR spectrum to the quenching efficiency.[32] Results show that the quenching efficiency increases as LSP band gets closer to the emission peak of the QDs, which is consistent with the FRET/SET theory that the transfer efficiency is directly dependent on the spectral overlap of the donor's emission with the acceptor's absorption. The quenching efficiency is around 98% for the GNRs-QDs pair whose LRS and emission spectrum overlaps best. There are studies on energy transfer process based on smaller GNRs (1.5nm in diameter) by linking nanorods with dye molecules with DNA strand. By varying the strand length, one can quantitatively control the distance between donor and acceptor. The quenching behavior is found to follow a d^4 relationship [31].

Surface plasmon plays an important role in energy transfer process. FRET is found to be enhanced in existence of metal particles.[33] The enhancement of Förster transfer occurs due to the effect of plasmon-assisted amplification of electric fields inside the nanoscale assembly. Simultaneously, metal nanocrystals lead to an increase of energy losses during the Förster transfer process. Govrov et.al have published theory discussion on enhanced FRET between semiconductor and metal nanopartilces.[59] Our group have demonstrated energy transfer effect between DAPI, a commonly used DNA dye, and gold nanoparticles under two-photon excitation in solution phase using fluorescence lifetime imaging microscopy (FLIM). With comparable size and concentration, gold nanorods (GNRs) provide more efficient energy transfer than gold nanospheres (GNSs). A dramatic reduction in fluorescence lifetime of DAPI from 1.7ns to less than 1ns was observed when 1.4ml of GNRs was added to a 2ml DAPI solution, indicating a strong energy transfer from DAPI to GNRs. While no change in the lifetime of DAPI was observed when same amount of GNS was added to the same DAPI solution. This difference between the transfer efficiency of GNRs and GNSs cannot be explained by either FRET or SET model. In the FRET model, the Förster distance (R_0), the separation at the transfer rate equals to the donor decay rate, are comparable for both rods and spheres, as the absorption spectrum of these two nanoparticles show similar structure in the DAPI emission range. While in the SET model, d_0, (the critical distance as R_0), are identical for GNRs and GNSs, as the parameters related to acceptor are derived from bulk gold. This large discrepancy is unlikely to be due to surface (size) effect, as the sizes of the two nanoparticles are comparable and both are significantly larger than dye molecule. We attribute this transfer enhancement effect to the longitudinal surface plasmon mode of GNRs overlapping with the incidence excitation wavelength. Thus the energy transfer would provide more detailed information in biological studies using GNRs as fluorescence probes, especially when combined with the advantages of two-photon excitation microscopy and more intense TPL from GNRs. GNRs would seem to have potential not only in FRET imaging, but also in energy transfer-based fluorescence lifetime sensing of bio-analytes such as metal ions and protein. [60].

4. Fluorescence Enhancement

According to radiating plasmon (RP) model for metal enhanced fluorescence (MEF), the scattering component of metal particle extinction contributes to MEF and the absorption component contributes to quenching. For gold colloids above 80 nm in diameter the light scattering yield is over 50% [61, 62], which suggests that MEF may occur.[34] Fluorescence from fluorophores in adjacent to gold nanoparticles can be enhanced. Moreover, the presence of a nearby metallic nanoparticle can not only enhance the quantum yield but also stabilize adjacent fluorophores against photobleaching. The fluorescence enhancement is attributable to enhanced absorption by the fluorophore, modification of the radiative decay rate of the molecule, and enhanced coupling efficiency of the fluorescent emission to the far field.[63] In 2007, Lakowicz's group reported gold nanoparticles enhanced fluorescence. They applied Alexa Fluor anti rabbit immunogolobulin G (IgG) onto gold films prepared by vapor deposition. Fluorescence intensity was found to be enhanced about 10-fold at maximum.[34] Bardhan et al. have demonstrated the quantum yield of dye IR800 increases from 7% to 84% and 74% by conjugating dye molecules to gold nanospheres and nanorods respectively. The enhancement effect is explained by both the scattering and absorption enhancement. GNRs predominantly enhance the emission of the fluorophore by absorption enhancement, owing to the high intensity near field resulting from the longitudinal plasmon resonance. [63].

5. Spaser

Spaser is short for surface plasmon amplification by stimulated emission of radiation. Pointed our by Dr. Stockman and Bergman, who firstly proposed the idea of spaser, it is analogous to the conventional laser for three important reasons: first, surface plasmons are bosons which have spin 1 just as photons do; second, surface plasmons are electrically neutral excitations; third, surface plasmons are the most collective material oscillations known in nature.[35] In a spaser photons are replaced by a nanoparticle supporting the plasmonic modes. The mechanism of spasering can be illustrated by the following structure: a noble nanoshell surrounded by a few monolayers of quantum dots (NQDs), which could be a possible design of a practical spaser. At first, an electron–hole pair is created by external excitation field and relaxes to an excitonic state due to carrier multiplication. Then the excitons would recombine to form photons. But in existence of the resonant localized surface plasmon, the excitonic energy is transferred to the resonant SPs of the metal nanoshell without emitting photons. The SPs stimulate further transitions in the gain medium, leading to the excitation of more identical SPs in the same SP mode, driving the action of the spaser. The most unique advantage of spaser compared to traditional optical sources is that it can generate dark modes, which means that a spaser will create highly intensive, localized field without any background radiation and radiative losses. If realized, a spaser can be applied to a variety of research fields ranging from sensing and biomedicine to imaging and information technology.[35] In 2009, Noginov and co-workers first demonstrated a spaser-based nanolaser by employing a high-brightness luminescent core–shell silica nanoparticles known as Cornell dots which composed of a gold core, providing plasmon modes, surrounded by a silica shell coated with organic dye Oregon Green 488 (OG-488), acting as a gain medium. The diameter of the gold core was around 14nm and the thickness of silica shell was around 15nm. The surface plasmon band of gold core centeres at 520nm. The number of dye molecules per nanoparticle is estimated to be 2.7×10^3 and calculated Q-factor is 14.8. Stimulated emission at 531nm was observed, and confirmed by both time-resolved and pump threshold experiments.[64].

SUMMARY

We have shown some of the examples of using GNPs in biological imaging and sensing based on their unique localized surface plasmon features. Characteristic short lifetime, decent intensity, together with polarization of TPL from gold nanorods, can be a promising contrast agent for use in luminescence microscopy and provide information on both biological system and environmental conditions. Furthermore, GNRs can be applied in both energy transfer and fluorescence enhancement research, and LSP play a critical role in these processes. By developing spaser technology, GNPs can be even more plausible in LSP study and application. In the near future, GNRs would provide opportunities to further study the surface plasmon effect at nanometer scale and drive the research on single molecules study, nanotechnology therapy, drug design, complex bio-system as well as new-generation photonic materials.

REFERENCES

[1] Pipino, A. C. R.; Schatz, G. C.; Van Duyne, R. P. *Phys. Rev. B* 1994, 49, 8320-8330.

[2] Geddes, C. D.; Aslan, K.; Gryzynski, I; Malicka, J; Lakowicz, J. R. *Rev. Fluoresc.* 2004, 104, 4313-4323.

[3] Jensen, T. R.; Van Duyne, R. P.; Johnson, S. A. Maoni, V. A. *Appl. Spectrosc.* 2000, 54, 371-377.

[4] Shimizu, K. T.; Woo, W. K.; Fisher, B. R.; Eisler, H. J.; Bawendi, M. G. *Phys. Rev. Lett*. 2002, 89, 117401.

[5] Haller, K. L.; Bumm, L. A.; Altkorn, R. I.; Zeman, E. J.; Schatz, G. C.; Van Duyne, R. P. *J. Chem. Phys.* 1989, 90, 1237-1252.

[6] Goulet, P. J. G.; Aroca, J. R. F. *Top. Fluoresc. Spectrosc*. 2005, 8, 223-247.

[7] Chen, Y.; Preece, J. P.; Palmer, R. E. *Ann. N. Y. Acad. Sci.* 2008, 1130, 201-206.

[8] Haes, A. J.; Chang, L.; Klein, W. L.; Van Duyne, R. P. *J. Am. Chem. Soc.* 2005, 127, 2264-2271.

[9] Haes, A. J.; Hall, W. P.; Chang, L.; Klei, W. L.; Van Duyne, R. P. *Nano. Lett.* 2004, 4, 1029-1034.

[10] McFarland, A. D.; Van Duyne, R. P. *Nano. Lett*. 2003, 3, 1057-1062.

[11] Yonzon, C. R.; Stuart, D. A.; Zhang, X.; McFarland A. D.; Haynes, C. L.; Van Duyne, R. P., Talanta 2005, 67, 438-448.

[12] Jung, L. S.; Campbell, C. T.; Chinowsky, T. M.; Mar, M. N.; Yee, S. S. *Langmuir*, 1998, 14, 5636-5648.

[13] Endo, T.; Kerman, K.; Nagatani, N.; Takamura, Y.; Tamiya, E. *Anal. Chem.* 2005, 77, 6976-6984.

[14] Brockman, J. M.; Nelson, B. P.; Corn, R. M. *Annu. Rev. Phys. Chem.* 2000, 51, 41-63.

[15] Haes, A. J.; Stuart. D. A.; Nie, S. M.; Van Duyne, R. P. *J. Fluoresc.* 2004, 14, 335-367.

[16] Smith. E. A.; Corn, R. M. *Appl. Spectrosc*. 2003, 57, A320-322.

[17] Srituravanich, W.; Fang, N.; Sun, C.; Luo, Q.; Zhang, X. *Nano Lett.* 2004, 4, 1085-1088.

[18] Kik, P. G.; Maier, S. A.; Atwater, H. A. *SPIE*, 2004, 5347, 215-223.

[19] Sundaramurthy, A. Schuck, P. J.; Conley, N. R.; Fromm, D. P.; Kino, G. S.; Moerner, W. E. *Nano. Lett*. 2006, 6, 355-360.

[20] Katherine, A. W.; Van Duyne, R. P. *Annu. Rev. Phys. Chem*. 2007, 58, 267-297.

[21] Maier, S. A.; Brongersma, M. L.; Kik, P. G., Meltzer, S.; Requicha, A. A. G. Atwater, H. A. *Adv. Mater*. 2001, 13, 1501-1505.

[22] Barnes, W. L.; Dereux, A.; Ebbse, T. W. Nature, 2003, 424, 824-830.

[23] Hutter, E.; Fendler, J. H. *Adv. Mater*. 2004, 16, 1685-1706.

[24] Emory, S. R.; Nie. S. *Science*, 1997, 275, 1102-1106.

[25] Haynes, C. I.; Yonzon, C. R.; Zhang, X. *J. Raman Spectrosc*. 2005, 36, 471-84.

[26] Yonzon, C. R.; Haynes. C. I.; Zhang, X.; Walsh, J. T.; Van Suyne, R. P. *Anal. Chem.* 2004, 76, 78-85.

[27] Huff, T. B.; Tong, L.; Zhao, Y.; Hansen, M. N.; Cheng, J-X.; Wei, A. *Nanomedicine* 2007, 2, 125-132.

[28] Bouhelier, A.; Bachelot, R.; Lerondel, G.; Kostcheev, S.; Royer, P.; Wiederrecht, G. P. *Phys. Rev. Lett.* 2005, 95, 267405.

[29] Wang, H.; Huff, T. B.; Zweifel, D. A.; He, W.; Low, P. S.; Wei, A.; Cheng, J-X. *Proc. Natl. Acad. Sci*. 2005, 102(44), 15752-15756.

[30] Sen. T.; Sadhu, S.; Patra, A. *Appl. Phys. Lett*. 2007, 91, 043104.

[31] Jennings, T. L.; Singh, M. P.; Strouse, G. F. *J. Am. Chem. Soc.* 2006, 128, 5462-5467.

[32] Li, X.; Qian, J.; Jiang, L.; He, S. *Appl. Phys. Lett*. 2009, 94, 063111.

[33] Zhang, J.; Fu, Y.; Chowdhury, M. H.; Lakowicz, J. R. *J. Phys. Chem. C* 2007, 111, 11784-11792.

[34] Zhang, J. Lakowicz, J. R. *Opt. Express* 2007, 15(5), 2598-2606.

[35] Stockman, M. I. *Nature Photon*. 2008, 2, 327-329.
[36] Zheludev, N. I.; Prosvirnin, S. L.; Papasimakis, N.; Fedotov, V. A. *Nature Photonics*, 2008, 351-354.
[37] Bergman, D. J.; Stockman, M. L. *Phys. Rev. Lett*. 2003, 90, 027402.
[38] Bruda, C.; Chen, X.; Narayanan, R.; El-Sayed, M. A. *Chem. Rev*. 2005, 105, 1025-1102.
[39] Murphy, C. J.; Gole, A. M.; Stone, J. W.; Sisco, P. N.; Alkilany, A. M.; Goldsmith, E. C.; Baxter, S. C. 2008, 41, 1721-1730.
[40] Mooradian, A. *Phys. Rev. Lett*. 1969, 22, 185-187.
[41] Boyd, G. T.; Yu, Z. H.; Shen, Y. R. *Phys. Rev. B*, 1986, 33(12), 7923-7936.
[42] Boyd, G. T.; Rasing, T.; Leite, J. R. R.; Shen, Y. R. *Phys. Rev. B*, 1984, 30(2), 519-526.
[43] Wang, D-S.; Hsu, F-Y.; Lin, C-W. *Optics Express* 2009, 17(14), 11350-11359.
[44] Tong, L.; Wei, Q.; Wei, A.; Cheng, J-X. *Photochem. Photobiol*. 2009, 85, 21-32.
[45] Imura, K.; Nagahara, T.; Okamoto, H. *J. Phys. Chem*. B 2005, 109, 13214-13220.
[46] El-Sayed, M. A. *Acc. Chem. Res*. 2001, 34, 257-264.
[47] Noguez, C. *J. Phys. Chem. B*. 2007, 111, 3806-3819.
[48] Varnavski, O. P.; Mohamed, M. B.; El-Sayed, M. A. *J. Phys. Chem. B* 2003, 107, 3101-3104.
[49] Biagioni, P.; Celebrano, M.; Grancini, G.; Brida, D.; Mátéfi-Tempfli, S.; Duò, L.; Hecht, B.; Cerullo, G.; Finazzi, M. *Phys. Rev. B* 2009, 80, 045411.
[50] Zhang, Y.; Yu, J.; Birch, D. J. S.; Chen, Y. Proc. SPIE 2011, 7910_51.
[51] Takahashi, H.; Niidome, Y.; Yamada, S. *Chem. Commun*. 2005, 17, 2247-2249.
[52] Hilgenbrink, A. R.; Low, P. S. *J. Pharm. Sci*. 2005, 94, 2135-2146.
[53] Sudimack, J.; Lee, R. J. *Adv. Frug Deliv. Rev*. 2000, 41, 147-162.
[54] Niidome, T.; Yamagata, M.; Okamoto, Y.; Akiyama, Y.; Takahashi, H.; Kawano, T.; Katayama, Y.; Niidome, Y. *J. Control. Release* 2006, 114, 343-347.
[55] Tong, L.; Zhao, Y.; Huff, T. B.; Hansen, M. N.; Wei, A. *Adv. Mater*. 2007, 19, 3136-3141.
[56] Huff, T. B.; Hansen, M. N.; Zhao, Y.; Cheng, J-X.; Wei, A. *Langmuir* 2007, 23, 1596-1599.
[57] Bestvater, F.; Spiess, E.; Stobrawa, G.; hacker, M.; Feuer, T.; Porwol, T.; Berchner-Pfannschmidt, U.; Wotzlaw, C.; Acker, H. *J. Micros*. 2002, 208, 108-115.
[58] Zhang, Y.; Yu, J.; Birch, D. J. S.; Chen, Y. *J. Biomed. Opt*. 2010, 15, 020504-3.
[59] Govorov, A. O.; Lee, J.; Kotov, N. A. *Phys, Rev. B* 2007, 76, 125308.
[60] Zhang, Y.; Birch, D. J. S.; Chen, Y. *Appl. Phys. Lett*. 2011, in press.
[61] Yguerabide, J.; Yguerabide, E. E. *Anal. Biochem*. 1998, 262, 137-156.
[62] Yguerabide, J.; Yguerabide, E. E. *Anal. Biochem*. 1998, 262, 157-176.
[63] Bardhan, R.; Grady, N. K.; Cole, J. R.; Joshi, A.; Halas, N. *Asc Nano* 2009, 3(3), 744-752.
[64] Noginov, M. A.; Zhu, G.; Belgrave, A. M.; Bakker, R.; Shalaev, V. M.; Narimanov, E. E.; Herz, E.; Stuteewong, T.; Wiesner, U. *Nature* 2009, 460, 1110-1112.

In: Plasmons: Structure, Properties and Applications
Editors: Anton E. Turunen and Johannes O. Niemi
ISBN: 978-1-62100-060-0

Chapter 6

STUDYING LIVING CELLS BY INFRARED SURFACE PLASMON SPECTROSCOPY

***Victor Yashunsky*[1*], *Alexander Zilberstein*[1], *Tal Marciano*[3], *Vladislav Lirtsman*[1], *Michael Golosovsky*[1], *Benjamin Aroeti*[2] and *Dan Davidov*[1]**

[1]The Racah Institute of Physics,
The Hebrew University of Jerusalem, Jerusalem, Israel
[2]Department of Cell and Animal Biology,
The Alexander Silberman Institute of Life Science,
The Hebrew University of Jerusalem, Jerusalem, Israel
[3]Laboratoire d'Optique Appliquée, ENSTA ParisTech-Ecole Polytechnique-CNRS, France

ABSTRACT

The development of new technologies capable of monitoring the delicate details of the dynamics of living cells, which remain outside the optical resolution, is vital for cell research. The infrared light introduces several unique advantages for cell studies. Its interaction with live tissue causes no photodamage and allows examining chemical information encoded in infrared absorption spectrum. This chapter presents newly developed infrared surface plasmon resonance (SPR) methodology. This chapter discusses a specific aspects related to studies of aqueous samples using infrared surface plasmon. The extreme sensitivity of the surface plasmon to refractive index changes together with extended probing depth in the infrared range in comparison to visible range allows studying living cell morphology with submicron precision. Analysis of the temporal variation of the SPR reflectivity allows reconstructing the dynamics of the cell shape and observing various stages in cell behavior. Using infrared SPR technique we have studied the dynamics of cellular processes and successfully monitored submicron changes in cell morphology.

---

* Corresponding author Electronic mail: victor.yashunsky@mail.huji.ac.il

INTRODUCTION

The imaging resolution of a conventional optical microscope is limited by diffraction to ~200 nm in the visible range [1]. Thus, delicate changes in living cell morphology stay out of reach of optical microscopy. The need to overcome such limits stimulated the development of alternative methods. Most existing methods involve labeling and tagging by fluorescent molecules, quantum dots, etc. Nevertheless, a number of label-free techniques have been developed over the past few years for studying cells *in vitro* [for reviews see: [2-7]].

Biosensors based on surface plasmon resonance (SPR) have been widely used for determination of the affinity and kinetics of a wide variety of molecular interactions in real time, without need for a molecular tag or label [8,9]. Recently, we have extended the conventional SPR method to the infrared wavelength range, using a Fourier transform infrared spectrometer (FTIR) as a light source [10]. Combination of infrared light with aqueous environment results in several unique advantages for living cell studies which are discussed in this chapter. The most important is the sampling depth of the surface plasmon (SP) which is wavelength-dependent and can reach up to few μm in infrared range [11]. This is sufficient for detection of processes occurring in large part of the cell volume whereas visible SPR-based techniques are limited to the events occurring in the vicinity of substrate [12].

We demonstrate that infrared surface plasmon spectroscopy can be applied to living cells and it is capable of monitoring the morphological changes occurring in individual cells or in entire tissue. We discuss effects of different molecular stimuli on the cellular activity.

METHODOLOGY

Surface Plasmon Resonance (SPR)

The surface plasmon is a transverse magnetic (TM) wave which propagates along the metal-dielectric interface (we denote it as *x*-direction) and decays in the direction perpendicular to the interface (*z*-direction). Surface plasmon (SP) can be resonantly excited by the electromagnetic wave impinging on the metal-dielectric interface using special couplers, such as prism or diffraction grating. Once SP has been excited, a sharp dip in the reflectivity spectrum appears. This dip indicates SPR. The resonant wavelength and the depth of the SPR minimum are extremely sensitive to any changes in refractive index of the dielectric layer on top of the metal. Existing SPR biosensors operate in visible range. Typically they are used to characterize the dynamics of molecular binding interactions through the change in the refractive index that occurs when molecules bind to the metal surface coated with antibodies [8,9].

The dispersion relation for a SP at metal-dielectric interface is:

$$k_{sp} = k'_{sp} + ik''_{sp} = \frac{2\pi}{\lambda}\left(\frac{\varepsilon_m \varepsilon_d}{\varepsilon_m + \varepsilon_d}\right)^{1/2} \tag{1}$$

The resonant wave vector is determined by the complex dielectric permittivities of the dielectric, $\varepsilon_d=(n_d+i\kappa_d)^2$, and of the metal, ε_m. Since, there is momentum mismatch $k'_{sp}>k_x=k_0 sin\Theta$, the SP on the metal-dielectric interface cannot be excited directly by an electromagnetic wave. However, several configurations allow overcoming this momentum mismatch. Figure 1B (upper panel) shows the dispersion curves for SP at Au-water interface excited using Kretschmann's geometry [13] based on ZnS prism (Figure 1A). SPR appears as a reflectivity dip, it occurs at the intersection of the dispersion curves for the surface plasmon and for the impinging wave, when SP momentum matches x-projection of the incident wave momentum, $k'_{sp}=k_x=n_{prism}k_0 sin\Theta$ (Figure 1B). The coupling angle, Θ is:

$$\Theta = \sin^{-1}\left[\frac{1}{n_{prism}}\left(\frac{\varepsilon'_m n_d^2}{\varepsilon'_m + n_d^2}\right)^{1/2}\right] \qquad (2)$$

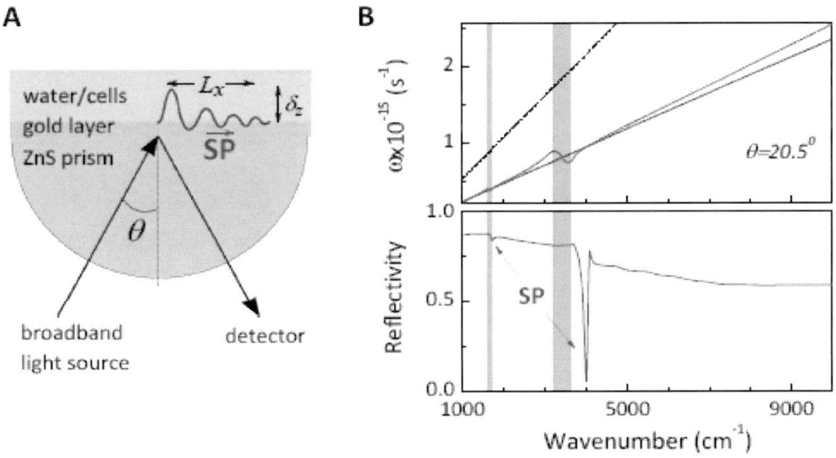

Figure 1. (A) Coupling scheme of surface plasmon excitation in the Kretschmann's geometry. (B) *Upper panel*: Dispersion relations for the surface plasmon on Au/water interface (red line) and for the light propagation inside ZnS prism (blue line). Dashed line corresponds to the light propagation in free space and weak blue bars denote major water absorption bands. *Lower panel*: Reflectivity spectrum for TM-polarized beam calculated for the tri-layer system, as it shown in Fig. 1A. Surface plasmon resonance appears as sharp reflectivity dip at ν=4000 cm^{-1} and ν=1725 cm^{-1}, when the momentum matching condition is satisfied and light energy is transferred to the surface plasmon.

The coupling angle is determined by the dielectric permittivities of the dielectric, $\varepsilon'_d(\lambda)\approx n_d^2$ $(\kappa_d<<1)$ and of the metal, $\varepsilon'_m(\lambda)$. Equation 2 shows that the refractive index of the analyte n_d for given wavelength can be found from angular or wavelength interrogation of the surface plasmon resonance. For wavelength interrogation, the resonant wavelength (λ_{sp}) is inexplicitly given by:

$$n_d(\lambda_{sp}) = n_{prism}\sin\Theta\left(\frac{|\varepsilon_m|}{|\varepsilon_m| + n_{prism}^2\sin^2\Theta}\right)^{1/2} \qquad (2a)$$

Note that n_{prism} and ε_m are wavelength-dependent, yet the dispersion of dielectric, $n_d(\lambda)$, is more significant due to existence of absorption bands in infrared wavelengths range.

Figures 2A and 2B demonstrate the angular interrogation measurements for infrared spectrum (1-10 μm wavelength or 10^5-10^4 cm^{-1} wavenumber). From these 2D reflectivity maps the spectrum of the refractive index of the analyte (n_d) was calculated (Figure 3A).

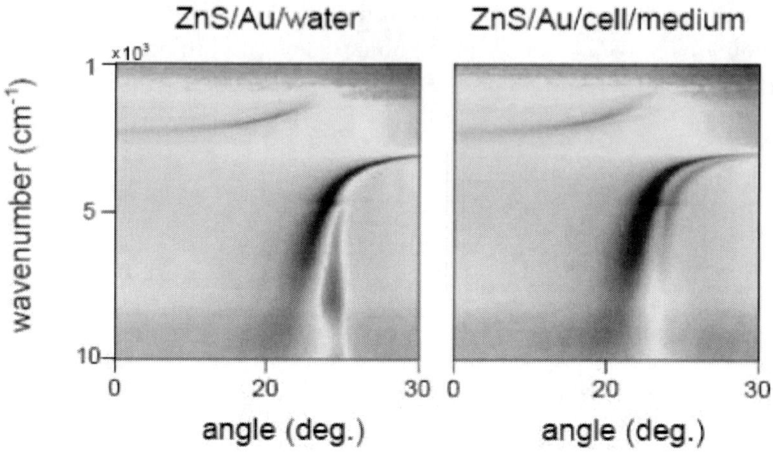

Figure 2. Reflectivity maps, $R(v,\theta)$ obtained for tri-layer (ZnS/Au/water) system (*left panel*) and four-layer (ZnS/Au/cells/medium) system (*right panel*). The intense blue stripes correspond to the surface plasmon resonance while the weak blue satellite (~22 deg., ~4500 cm^{-1}) in the right panel corresponds to the guided mode resonance. The narrow yellow stripe in between indicates the critical angle. The wavelength dependence of the resonances arises mostly from the water dispersion. The discontinuity at 3300 cm^{-1} arises from the water absorption band.

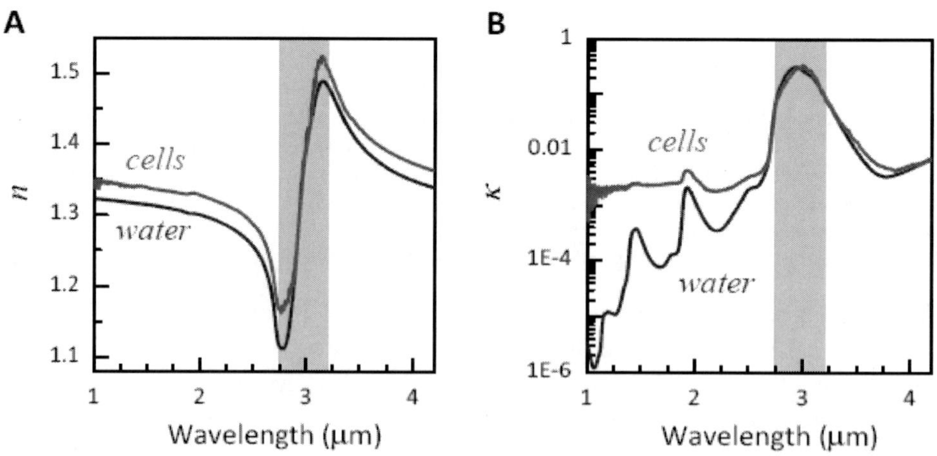

Figure 3. Optical constants of water (blue line) and 10 μm thick MDCK cell monolayer (red line) were derived from the reflectivity measurements. The real part of the cell refractive index, n (*left panel*) slightly exceeds that of the water. The imaginary part of refractive index, κ (*right panel*) is determined by absorption and scattering. While for the water the main contribution to κ comes from the absorption, for the cells scattering becomes more significant, especially at shorter wavelengths.

Losses

Analysis of the resonant reflectivity allows measuring of the lossy part of the analyte refractive index (κ_d) that is determined by absorption and scattering. Reflectivity nearby the SPR can be approximated by Lorentzian [13]:

$$R = \left|r_{mp}\right|^2 \left(1 - \frac{4k''_{sp}\Gamma_{rad}}{(k_x - k'_{sp})^2 + (k''_{sp} + \Gamma_{rad})^2}\right) \quad (4)$$

Here, r_{mp} is the Fresnel reflection coefficient (1) at the prism-metal interface, and Γ_{rad} is radiation loss of the SP caused by finite thickness of the metal layer. The resonant reflectivity is:

$$R_{\min} = \left| r_{mp} \frac{k''_{sp} - \Gamma_{rad}}{k''_{sp} + \Gamma_{rad}} \right|^2 \quad (5)$$

The reflectivity minimum (R_{min}) is determined by the k''_{sp} which depends on losses in the dielectric layer and in the metal film, ε''_d and ε''_m,

$$k''_{sp} = k_0 \left(\frac{\varepsilon'_m \varepsilon'_d}{\varepsilon'_m + \varepsilon'_d}\right)^{3/2} \left(\frac{\varepsilon''_d}{\varepsilon'^2_d} + \frac{\varepsilon''_m}{\varepsilon'^2_m}\right) \quad (6)$$

From the R_{min} we were able to assess the imaginary part of the analyte refractive index, $\kappa_d = \varepsilon''_d/2n_d$ (Eqs. 5 and 6). Figure 4B shows κ_d for water and cells as determined from the corresponding reflectivity measurements (Figures 2A and 2B).

Sampling Depth

The intensity of the SP field in the analyte exponentially decays along z-direction. For sensing applications the sampling depth of the SP (δ_z) is defined from the exponential decay of the intensity, $I(z)=|E_{(z=0)}|^2 \exp(-z/\delta_z)$ [12] and it depends on the wavelength of the exciting wave (Figure 4A),

$$\delta_z = \mathrm{Im}\left[(\varepsilon_m + \varepsilon_d)^{1/2} \lambda/(4\pi\varepsilon_d)\right] \quad (7)$$

When properties of the analyte layer vary in z-direction, the complex refractive index sensed by SP is given by the integral:

$$n_d + i\kappa_d = \frac{1}{\delta_z} \int_{z=0}^{z=\infty} [n(z) + i\kappa(z)] \exp(-z/\delta_z) dz \quad (8)$$

For cell layer with thickness d_{cell} immersed into the growth medium, Eq. 8 reduces to,

$$n_d + i\kappa_d = (n_{medium} + i\kappa_{medium}) + (\Delta n + i\Delta\kappa)\exp(-d_{cell}/\delta_z)$$

where $\Delta n = n_{cell} - n_{water}$ and $\Delta\kappa = \kappa_{cell} - \kappa_{water}$.

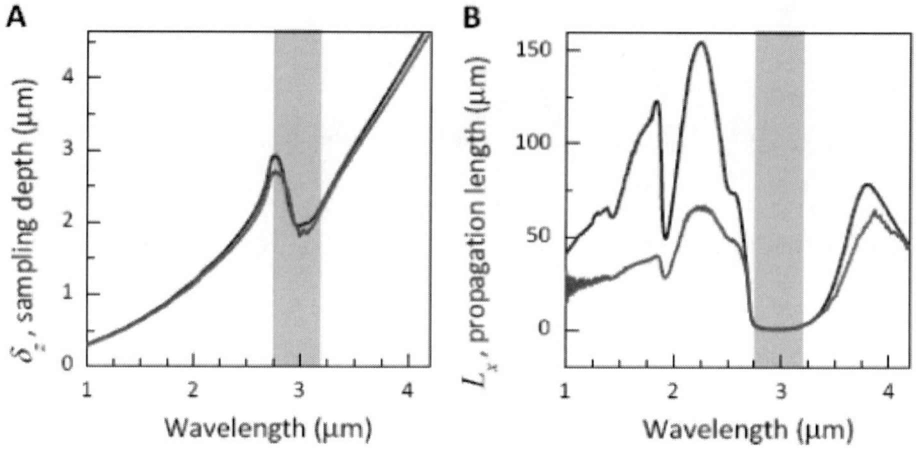

Figure 4. Sampling depth (A) and propagation length (B) for surface plasmon at Au/water and at Au/cell interfaces calculated from reflectivity spectra Figs. 2A and 2B.

Spatial Resolution

A crucial requirement for materials used in biological experiments with living cells is absence of toxic effects on cells. On the other hand, the conditions $\varepsilon'_m<0$, $\varepsilon''_m <|\varepsilon'_m|$ and $\varepsilon'_m> \varepsilon'_d$ have to be satisfied to observe SP. This requirement excludes the most part of the metals for the surface plasmon bio-application and leaves the Au as a metal coating. For functionality of living cells, the water-based media should be used. At these conditions, the SP propagation length $L_x=1/2k''_{sp}$ [12], is shown in Figure 4B. For a discontinuous dielectric layer there are two possible regimes: $L_x>d$ or $L_x<d$ where d is the characteristic length of the continuous region (e.g. ~20 μm for eukaryotic cell, ~1 μm for bacteria). In the first case there appears a single SPR dip corresponding to some effective medium [11], but in the second case few SPR dips at different wavelengths (at a fixed angle) can be observed (see RESULTS, section *D*).

EXPERIMENTAL

Setup

The experimental setup is schematically depicted in Figure 5A; the right-angle ZnS prism (20x40 mm^2 base; ISP Optics, Irvington, NY) coated by a thin Au film (d=10-20-nm, prepared by electron-beam evaporation) is attached to a 2 mL flow chamber filled with cell medium. The temperature of the flow chamber was set at 37°C and the flow rate was controlled via a motorized bee syringe pump.

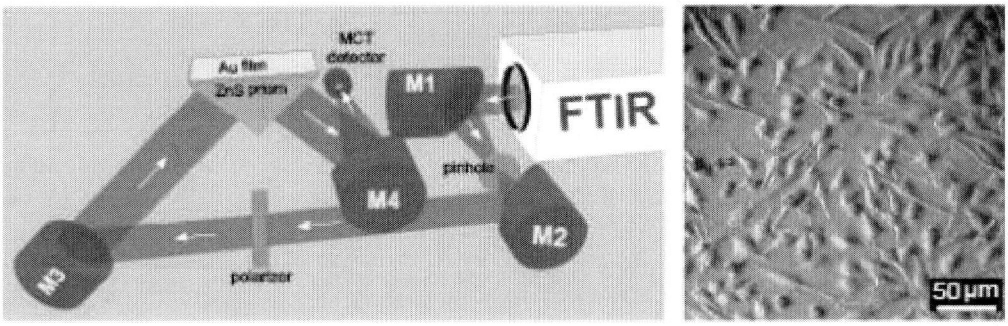

Figure 5. (A) Schematic description of the experimental setup for SPR excitation in the infrared range. The divergent wideband infrared beam (1-10 μm) is focused on the pinhole by the parabolic mirror (M1), collimated by another parabolic mirror M2 and polarized. Then it is reflected from the Au-coated prism in contact with analyte and focused on the MCT detector. (B) The image of HeLa cells cultured on ZnS/Au coated prism.

The prism- flow-cell assembly was mounted on the vertical translation stage. The FTIR spectrometer (Equinox 55, Bruker) was used as the broadband infrared source for SP excitation. The p-polarized infrared beam was collimated and reflected from the Au-coated prism base, and then focused onto a liquid-nitrogen-cooled MCT (HgCdTe) detector. The measurement of each FTIR spectrum lasted 25 s and represents an average over eight scans, with 8 cm^{-1} resolution. The s-polarized reflectivity spectrum was used as a background. Angular interrogation was performed in a slightly different geometry using Θ-2Θ motorized stage and vertical arrangement of the reflecting surface [10].

Cell Preparation

HeLa (cervical carcinoma cells), MEL 1106 (human melanoma) and A431 (epidermoid carcinoma cells) cell lines were cultured in 10-cm diameter dishes in CO_2 incubator (5% CO_2, 37°C, 90% humidity) using a growth medium, Dolbecco's modified Eagle's medium (DMEM, 10% (v/v) fetal calf serum, 1% (v/v) antibiotics). Madin-Darby canine kidney cells (MDCK) were cultured in minimal essential growth medium (MEM, 5% (v/v) fetal calf serum and 1% (v/v) antibiotics.

For SPR experiments the cells were prepared as follows: cells monolayer was detached from the dish by trypsin solution (0.05% or 0.25% trypsin/EDTA in Puck's saline A) and resuspended in 10mL of growth medium. Cell suspension was seeded on top of the Au-coated ZnS prism and placed inside the CO_2 incubator for 3-5 days until continuous monolayer was formed. The SPR measurements were performed within Hepes containing medium (MEM supplemented with 20 mM Hepes, pH 7.5). Other supplementations for MEM-Hepes medium are discussed in the RESULTS section.

LIVING CELL STUDIES, RESULTS AND DISCUSSION

A. Cell Adhesion, Spreading and Detachment

Using infrared SPR spectroscopy we studied the process of epithelia monolayer formation from suspended cells. MDCK cells suspended in MEM-Hepes medium were injected into the flow chamber. After 15 min the suspended cells descend to the Au substrate. Then we established the cell-free medium flow. Cells were left to adhere until they formed a contiguous monolayer (~4 hours). Finally, the cells were detached from the substrate using trypsin solution.

A qualitative assessment of this process was done by optical microscopy through the optical window on top of the flow cell [14]. Simultaneous SPR measurements revealed different stages of cell-substrate and cell-cell attachment. Since the cell refractive index is slightly higher than refractive index of the cell medium, $\Delta n = n_{cell} - n_{medium} \approx 0.03$, then the dynamics of the SPR wavenumber ($v = 1/\lambda_{sp}$) follows the refractive index change nearby the Au substrate (Eq. 2a). The SPR red shift is proportional to the cell confluence resulting from cell spreading on the substrate, $\Delta v = S(v_{sp})f$. Indeed, during cell spreading the resonant wavenumber is monotonously red-shifted (Figure 6A, lower panel). On the other hand the SPR reflectivity (Figure 6B, upper panel) changes are mostly influenced by the amount of cell-medium boundaries. These boundaries have major contribution to k''_{sp} and consequently to R_{min} (Eq. 6).

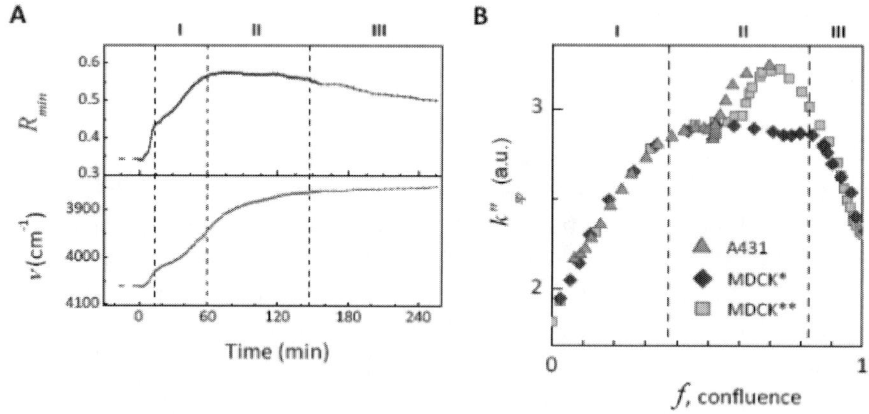

Figure 6. (A) Time dependence of the surface plasmon wavenumber v and depth, R_{min} during cell deposition and spreading. t=0 corresponds to cell injection. *Lower panel*: the SPR is continuously shifted indicating increase of the refractive index of the layer contacting the prism. This is associated with cell deposition. *Upper panel*: the surface plasmon depth (which is determined by the scattering on cells) varies nonmonotonously through the cell deposition process. Dashed lines separate between different stages defined in Figure 6B. (B) Dependence of the k''_{sp} on the surface coverage f (estimated from R_{min} and Δv) during cell deposition for A431 cells in serum-free medium (orange triangles), MDCK cells in serum-free medium (purple rhombs), and MDCK cells in 10% (v/v) fetal calf serum medium (grey squares). This graph encodes the information on cell morphology for different processes; individual cell spreading (stage I), cell clustering (stage II), continuous cell monolayer formation (stage III).

After 60 min of spreading when the cell confluence gets close to 30%, the neighboring cells start to attach to each other (stage II) and the number of cell-media boundaries decreases whereas cell coverage increases. It means that by measuring R_{min} we follow the kinetics of cell-cell attachment.

Time excluding representation, k''_{sp} vs. f, is shown in Figure 6B. It yields a map where the trajectory of the cell spreading process depends only on the morphology. Cell spreading or detachment can differ not only by their rate, but also by their geometry. A good example is the development of protrusions (lamellipodia) associated with cell motility during merging of cell clusters (stage II). Indeed, the MDCK cells in the absence of growth factors in the medium, the MDCK cells in the presence of growth factors, and A431 cells without growth factors show very different trajectories in this representation. The first result is consistent with the fact that growth factors are stimulating actin polymerization required for protrusion formation [15]. In the second case, malignant A431 cells are much more mobile than MDCK cells, which correlates with the enhanced protrusion formation at stage II [16].

In the process of cell detachment a proteolytic enzyme (trypsin) breaksdown the protein bonds and dissociates adherent cells from each other and from the substrate. Cell detachment induced by trypsin solution occurs much faster than cell attachment to substrate and has an opposite tendency than the cell spreading. Trypsin disrupts the adhesion sites (i.e. cell-substrate and cell-cell junctions) and thus induces cell contraction. As a result cells adopt a spherical shape. This change is monitored by the SPR wavenumber blue-shift (Figure 7, lower panel) indicating decrease of refractive index, n_d, while the reflectivity drop (Figure 7, upper panel) indicating decrease of cell-media boundaries due to detachment of the cells from the substrate.

In this part we have demonstrated that infrared SPR spectroscopy allows not only following cell attachment/detachment process, but it is also capable of providing information about the cell morphology during different stages.

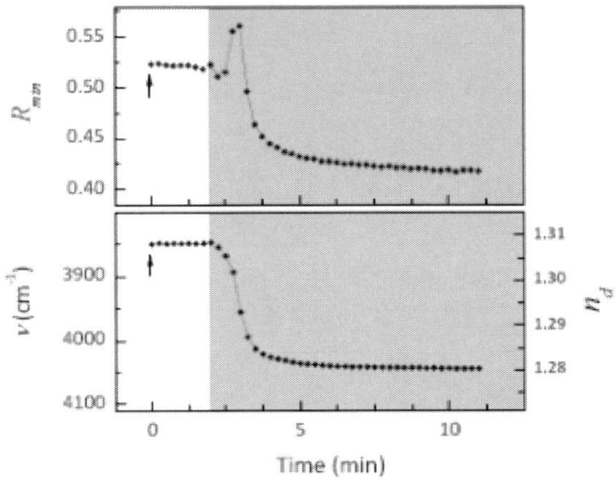

Figure 7. Dynamics of the cell detachment under the trypsin treatment. MDCK cell monolayer was treated by trypsin solution (0.25% (vol/vol)). Introduction of trypsin enzyme into extracellular medium (indicated by the arrow) leads to dissociation of adherent cells from each other and from the substrate. The detachment from the substrate (blue region) appears as the blue-shift of the SPR (*lower panel*), and the drop of the reflectivity at SPR (*upper panel*). The short jump in the reflectivity at 3 min is associated with initial stage of cell-cell dissociation.

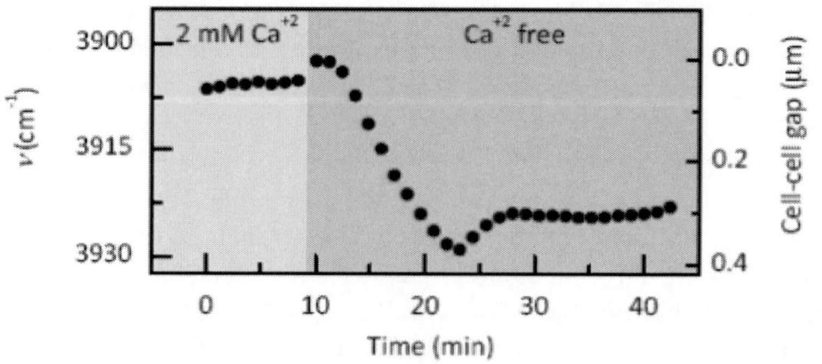

Figure 8. Dynamics of the SPR wavenumber following cell response to Ca^{+2}-free medium. Polarized MDCK monolayer was exposed to Ca^{+2}-free medium which leads to dissociation of cell-cell junctions, separation of adjacent cells, and to the opening of intercellular gaps (right *y*-scale). SPR blue shift initiated shortly after the injection of Ca^{+2}-free medium is associated with penetration of extracellular medium between the cells, and the corresponding reduction of the effective refractive index sensed by the surface plasmon wave.

B. Calcium Depletion

In this experiment we followed the response of polarized MDCK to decrease of free extracellular Ca^{2+}. For this purpose the MEM-Hepes medium was replaced by Ca^{2+}-free MEM-Hepes medium. Decrease of extracellular Ca^{2+} concentration leads to a reduction in the homophilic affinity between cadherins and decreases the adhesion between adjacent cells [17].

On the other hand, this affects the contraction applied by the cytoskeleton (i.e. actin filaments) separating the cells from each other, in such a way that the extracellular medium can penetrate between the cells [18]. Entrance of the medium is tracked by the blue-shit of the SPR (Figure 8). Using previously measured optical constants of cells and medium we estimate the cell-cell gap opened in between adjacent cells as

$$(cell\text{-}cell\ gap) \approx 2L_{cell}\,(1\text{-}f^{1/2})$$

Here L_{cell}~20 μm [18] denotes the typical cell length, f is the confluency measured from the SPR wavenumber shift, Δv (see *A.*) The maximal estimated opening of the cell gap is ~0.1 μm and it happenes after ~10 min. This result is consistent with previously reported measurements of transepithelial resistance variation under low extracellular Ca^{2+} conditions [19].

Monitoring of cell-cell attachment in living cells is a challenge since it involves spatial scales which are beyond optical resolution limit. The cell-cell attachment is usually studied by indirect methods (e.g. transepithelial resistance, penetration of small fluorescent molecules and functional tests) or on fixed cells through electron microscopy where the finest cell structures can be observed. However, first methods lack structural information and electron microscopy does not allow following the dynamics. Here, we have shown that infrared SPR spectroscopy is capable of monitoring submicron changes in cell shape in real-time which makes it an appropriate tool for cell adhesion studies.

C. Chlorpromazine Treatment

Chlorpromazine (CPZ), an amphipathic, antipsychotic agent, causes membrane bending in cells by modulation of the phospholipid bilayer. Such effect has been shown in previous studies for red blood cells and endothelial cells [20]. Here we investigated the effects of CPZ on the shape of confluent human melanoma (MEL 1106) monolayer. Exposure of cells to 10 μM of CPZ changes the curvature of the cell membrane, in particular that in contact with the substrate. Here, we monitor this morphological change by the SPR method (Figure 9A).

The SPR wavenumber tracks the total area covered by the cells (*see subsection* A.). The SPR reflectivity is proportional to the total perimeter of cell-medium boundaries inside the volume sensed by the SP. Exposure of the monolayer to CPZ induces immediate decrease of cell coverage and increase of the cell-medium boundaries (Figures 9A and 9B).

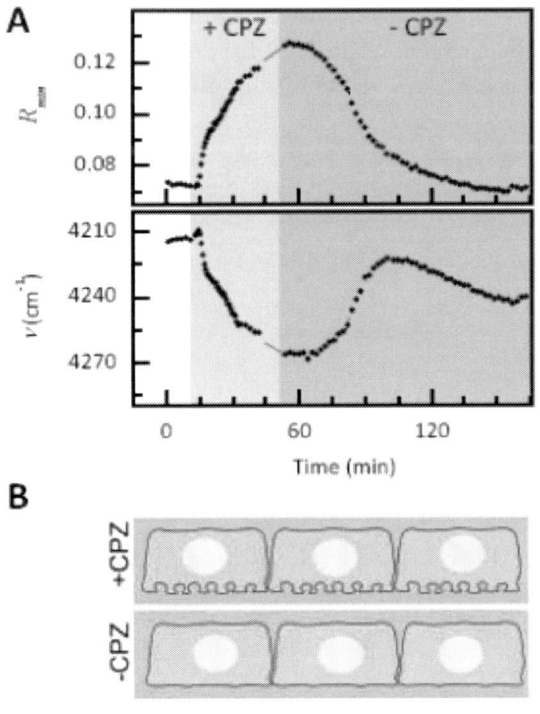

Figure 9. Human melanoma cells were treated with cationic amphiphilic drug chlorpromazine (CPZ 10 μM) which inhibits clathrin-mediated endocytosis and leads to cell deformation and to the increase of cell membrane curvature. (A) The reflectivity at SPR (*upper panel*) and the SPR wavenumber (*lower panel*) track the changes in cell shape. After CPZ was washed out from the extracellular medium, the reflectivity and the wavenumber return to their original values indicating restoration of the normal membrane shape. (B) Schematic illustration of the changes in cell shape under CPZ treatment. The upper panel shows the cell membrane deformation in the presence of CPZ (+CPZ), the lower panel shows the normal cell conformation in the absence of CPZ (-CPZ).

This response is somehow similar to cell trypsinization (A.) or low Ca^{2+} treatment (B.) where cells were detached from each other or from the substrate. Although in the case of the CPZ treatment same change in SPR signal is resulted from another mechanism. Here we monitor the increase in membrane curvature rather than cell detachment. Additional evidence to such curvature change is the short recovery time. After the removal of CPZ drug SPR parameters turns back to their original values.

This example emphasizes the potential of the infrared SPR spectroscopy for cell morphology studies. Although the cell response to CPZ was studied by electron microscopy, it revealed only qualitative information on membrane deformation. Using this previous knowledge [20] we successfully measured the dynamics of this process.

D. Cell Confluence

The following measurements demonstrate the spatial resolution of our infrared spectroscopic technique. As we showed in the METHODOLOGY, *Spatial resolution* section, the propagation of infrared surface plasmon on Au- cells/medium interface is wavelength-dependent, mostly due to water absorption. Adjustment of SPR wavenumber/wavelength allows observation of well separated resonances for different regions (i.e. the regions covered by cells and covered by medium) caused by the difference in the refractive index of the dielectric layer, $n_{medium} \neq n_{cell}$ $(\Delta n \approx 0.03)$. The depth of each resonance is determined by the ratio of cell-covered or uncovered area to the whole sensed area.

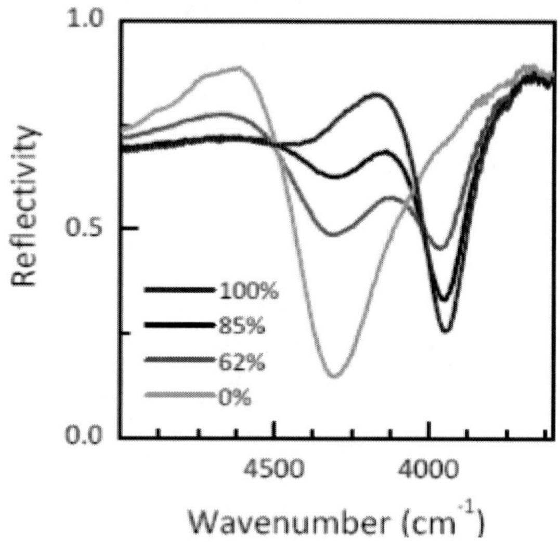

Figure 10. Reflectivity spectra for HeLa cells at various confluences. For uniform dielectric layer SPR appears as a single reflectivity dip (green and blue lines, zero and full coverage). However, for partial cell coverage there appear two reflectivity dips (4250 cm^{-1} and at 3950 cm^{-1}) arising from SPR on Au/medium and from SPR on Au/cell interface, respectively.

Figure 10 shows several SPR curves for various cell confluences in malignant epithelial cells (HeLa). For partial cell coverage, the SPRs appear as separate dips due to relatively small propagation length for given wavelength ($L_x \approx 50$ μm in cells, $L_x \approx 150$ μm in medium). Note, that HeLa cells are not equally distributed on the substrate, but form clusters containing several cells. Additional reason for two separate resonances is the spot size of impinging light beam. The measured spectrum is collected from the area of $\sim 10^7$ μm^2 while cell cluster area is approximately 10^4 μm^2. It means that each spectrum is a superposition of SPR spectra from regions with two different refractive indexes. Thus, the depths of each resonance correspond to the area of each region. This fact allows estimation of cell confluence as it shown in Figure 10.

CONCLUSION

In this chapter we have demonstrated a novel non-invasive and label-free methodology to adherent cells study. After a brief introduction to infrared surface plasmon spectroscopy, we showed several examples of how various cellular processes can be monitored using infrared surface plasmon spectroscopy. The infrared surface plasmon spectroscopy allows following of submicron variations in cell morphology which cannot be simply resolved by the optical microscopy. Although additional work is required to further understand the mechanism of surface plasmon propagation inside living cells, our findings confirm its value for cell screening.

REFERENCES

[1] Born, M. and E. Wolf. 1999. Principles of Optics: Electromagnetic Theory of Propagation, Interference and Diffraction of Light: Cambridge University Press. 986 p.

[2] Cooper, M. A. 2002. Optical biosensors in drug discovery. *Nature Review Drug Discovery* 1:515-528.

[3] Cooper, M. A. 2006. Non-optical screening platforms: the next wave in label-free screening? *Drug Discovery Today* 11:1068-1074.

[4] Giaever, I. and C. R. Keese. 1991. Micromotion of Mammalian-Cells Measured Electrically. *Proceedings of the National Academy of Sciences of the United States of America* 88:7896-7900.

[5] Hug, T. S., J. E. Prenosil, and M. Morbidelli. 2001. Optical waveguide lightmode spectroscopy as a new method to study adhesion of anchorage-dependent cells as an indicator of metabolic state. *Biosensors and Bioelectronics* 16:865-874.

[6] Ramsden, J. J., S. Y. Li, E. Heinzle, and J. E. Prenosil. 1995. Optical Method, for Measurement of Number and Shape of Attached Cells in Real-Time. *Cytometry* 19:97-102.

[7] Fang, Y., A. M. Ferrie, N. H. Fontaine, J. Mauro, and J. Balakrishnan. 2006. Resonant waveguide grating biosensor for living cell sensing. *Biophysical Journal* 91:1925-1940.

[8] Homola, J. 2003. Present and future of surface plasmon resonance biosensors. *Analytical and Bioanalytical Chemistry* 377:528-539.

[9] Knoll, W. 1998. Interfaces and thin films as seen by bound electromagnetic waves. *Annual Review of Physical Chemistry* 49:569-638.

[10] Lirtsman, V., M. Golosovsky, and D. Davidov. 2008. Infrared surface plasmon resonance technique for biological studies. *Journal of Applied Physics* 103:014702.

[11] Golosovsky, M., V. Lirtsman, V. Yashunsky, D. Davidov, and B. Aroeti. 2009. Midinfrared surface-plasmon resonance: A novel biophysical tool for studying living cells. *Journal of Applied Physics* 105:1020-1021.

[12] Johansen, K., H. Arwin, I. Lundstrom, and B. Liedberg. 2000. Imaging surface plasmon resonance sensor based on multiple wavelengths: Sensitivity considerations. *Review of Scientific Instruments* 71:3530-3538.

[13] Raether, H. 1988. Surface-Plasmons on Smooth and Rough Surfaces and on Gratings. *Springer Tracts in Modern Physics* 111:1-133.

[14] Yashunsky, V., V. Lirtsman, M. Golosovsky, D. Davidov, and B. Aroeti. 2010. Real-time monitoring of epithelial cell-cell and cell-substrate interactions by infrared surface plasmon spectroscopy. *Biophys J.* 99:4028-4036.

[15] Ridley, A. J. and A. Hall. 1992. The small GTP-binding protein rho regulates the assembly of focal adhesions and actin stress fibers in response to growth factors. *Cell* 70:389-399.

[16] Friedl, P. and K. Wolf. 2003. Tumour-cell invasion and migration: diversity and escape mechanisms. *Nat. Rev. Cancer* 3:362-374.

[17] Gumbiner, B. M. 1996. Cell adhesion: The molecular basis of tissue architecture and morphogenesis. *Cell* 84:345-357.

[18] Frixione, E., R. Lagunes, L. Ruiz, M. Urban, and R. M. Porter. 2001. Actin cytoskeleton role in the structural response of epithelial (MDCK) cells to low extracellular Ca2+. *J. Muscle Res. Cell Motil.* 22:229-242.

[19] Citi, S. 1992. Protein kinase inhibitors prevent junction dissociation induced by low extracellular calcium in MDCK epithelial cells. *J. Cell Biol.* 117:169-178.

[20] Hueck, I. S., H. G. Hollweg, G. W. Schmid-Schonbein, and G. M. Artmann. 2000. Chlorpromazine modulates the morphological macro- and microstructure of endothelial cells. *Am. J. Physiol. Cell Physiol.* 278:C873-878.

In: Plasmons: Structure, Properties and Applications
Editors: Anton E. Turunen and Johannes O. Niemi
ISBN: 978-1-62100-060-0

Chapter 7

INTERFACE PLASMON POLARITONS IN UNIAXIAL BIANISOTROPIC STRUCTURES

Roland H. Tarkhanyan[*]
Institute of Radiophysics and Electronics, 378410, Ashtarak-2, Armenia

ABSTRACT

A new class of nonradiative surface plasmon polaritons guided by the interface between two enantiomeric anisotropic chiral plasmonic media is presented. The waves are caused by bianisotropy and are absent in the case of enantiomorph bi-isotropic materials as well as in anisotropic structures without magnetoelectric coupling. The basic properties of the waves, in particular, the dispersion relations, localization conditions and propagation lengths differ dramatically from those for conventional surface plasmon polaritons propagating along the interface between different achiral media.

1. INTRODUCTION

Since the first publication by Sommerfeld in 1899 [1], surface electromagnetic waves became an object for enormous number of theoretical and experimental investigations (see, for example, refs. [2-11] and citations therein). Surface waves are unique tools to explore the properties of material interfaces and are widely used in different areas of science and technology, particularly for sensing physical, chemical and biological agents [12-14]. Owing to the recent rapid development in optics and applications of artificial metamaterials for realization of the superlensing phenomenon and negative refraction[15-21], there is now a strong interest in surface waves at the interfaces of various types of complex electromagnetic materials[18,22,23]. Among the complex media the bi-isotropic and bianisotropic materials, which exhibit right- or left-handedness and magnetoelectric coupling, are one of the most challenging substances because of the essential role of the cross polarizability effects in their electromagnetic behavior. Without any doubt, the design of new composites based on bi-

[*] Email: rolandtarkhanyan@yahoo.com

isotropic and bianisotropic materials is of considerable physical and applied interest too, because it opens a unique possibility to construct novel types of devices (for example, biosensors), where electromagnetic waves propagate in a nonconventional way [24-29].

The purpose of this chapter is to explore the nature of nonradiative surface plasmon polaritons guided by the planar interface of two enantiomeric bianisotropic media. These waves like well known Zenneck [30] and Fano [31] waves, as well as surface plasmons at optical frequencies [32], propagate parallel to the interface with amplitude that decays exponentially with distance from the interface. Usually surface electromagnetic waves occur at the planar interface of different materials (e.g., a conductor and a dielectric), which can be either optically isotropic or anisotropic. In the isotropic case, a special frequency dependence of the permittivity (or the permeability in the case of magnetic materials) is required for the interface wave to occur, and the solutions exist within definite frequency intervals. In particular, it is necessary that the permittivity (or permeability) of one of the contacted media be negative. In the case when at least one of the media is optically anisotropic, the existence of interface waves does not impose any conditions on the permittivity [33]. Such nondispersive, so called Dyakonov surface waves appear at arbitrary frequencies and recently have been observed experimentally [34]. The boundary–value problem for Dyakonov-like waves propagating along a planar interface of two structurally chiral nonmagnetic materials was formulated and numerically solved in Ref.[35], assuming that either of the materials is twisted arbitrarily with respect to the direction of propagation. The waves considered in [35] are result of anisotropy; they occur when at least one of the optical axes is parallel to the interface and exist even in the absence of the chirality. In recent work [36], the existence of nondispersive surface electromagnetic waves guided by an interface between two transparent enantiomeric bianisotropic media has been considered in the case when the optic axes of both media are perpendicular to the interface. Note that in this case the existence of the Dyakonov-like surface waves is impossible: they can only propagate in a certain range of angles with respect to the optic axis parallel to the interface (see Ref. [33]).

Unlike all the above mentioned waves, the focus of this work is on electromagnetic waves at the interface of two identical homogeneous bianisotropic plasmonic materials with opposite handedness and optical axes perpendicular to the interface plane. In other words, we will consider the waves caused by combined influence of chirality and anisotropy. It is worth noting that at the planar interface of bi-isotropic enantiomorph materials the propagation of nonradiative surface waves is impossible [36]. Therefore, investigated in this work surface plasmon polaritons can be considered as a novel class of surface electromagnetic waves in the case of uniaxially bianisotropic enantiomeric materials.

The chapter is organized in nine main sections. Section 2 describes material tensors and constitutive relations used for the analysis of the electromagnetic properties of the enantiomeric uniaxial media. Section 3 discusses general dispersion relation and the field structure of the partial waves in both media with right- and left-handedness. The dispersion relationships for the surface plasmon polaritons are obtained in section 4. In section 5, the frequency regions of the wave existence in the absence of dissipation are determined. In Section 6, the dispersion curves of lossless surface plasmon polaritons are examined. The effect of the wave dissipation on the refractive index is considered in Section 7. In Section 8, the propagation length of the surface plasmon polaritons in lossy structures is determined. The concluded remarks are highlighted in section 9,

2. MATERIAL TENSORS AND CONSTITUTIVE RELATIONS

Consider an enantiomeric bilayer consisting of uniaxial chiral plasmonic materials with opposite handedness. The interface is assumed to be the z=0 plane; the material with left handedness (L-material) occupies the half-space region z>0 and the region z<0 is taken to be enantiomorph material with right handedness (R-material). Optical axes in both materials are supposed to be perpendicular to the interface plane. In this simplest case, the relative permittivity and permeability tensors of both R and L materials are given by

$$\hat{\varepsilon} = \begin{pmatrix} \varepsilon_\perp & 0 & 0 \\ 0 & \varepsilon_\perp & 0 \\ 0 & 0 & \varepsilon_\parallel \end{pmatrix}, \quad \hat{\mu} = \begin{pmatrix} \mu_\perp & 0 & 0 \\ 0 & \mu_\perp & 0 \\ 0 & 0 & \mu_\parallel \end{pmatrix} \tag{1}$$

while the chirality admittance tensor

$$\hat{\xi}_L = \begin{pmatrix} \xi_\perp & 0 & 0 \\ 0 & \xi_\perp & 0 \\ 0 & 0 & \xi_\parallel \end{pmatrix} \tag{2a}$$

for the L-material and

$$\hat{\xi}_R = -\hat{\xi}_L \tag{2b}$$

for its mirror image. In the case of reciprocal media both $\xi_\perp$ and $\xi_\parallel$ are real and are assumed to be positive constants while the components $\varepsilon_\perp, \varepsilon_\parallel$ may have the different sign. In the following, we restrict ourselves to the consideration of lossless reciprocal nonmagnetic materials with $\mu_\perp = \mu_\parallel = 1$ and

$$\varepsilon_{\perp,\parallel}(\omega) = \varepsilon'_{\perp,\parallel}(\omega) + i\varepsilon''_{\perp,\parallel}(\omega). \tag{3a}$$

Using Drude free electron gas model and neglecting contributions from the bound charges, we can set [15]

$$\varepsilon'_{\perp,\parallel} = 1 - \frac{\omega^2_{p\perp,\parallel}\tau^2_{\perp,\parallel}}{1+\omega^2\tau^2_{\perp,\parallel}}, \quad \varepsilon''_{\perp,\parallel} = \frac{\omega^2_{p\perp,\parallel}\tau_{\perp,\parallel}}{\omega(1+\omega^2\tau^2_{\perp,\parallel})}, \tag{3b}$$

where $\omega_{p\perp,\parallel}^2 = Ne^2 / m_{\perp,\parallel}$, ω is the angular frequency of the waves, e is the charge and N is the density of the free charge carriers, $m_{\parallel,\perp}$ and $\tau_{\parallel,\perp}$ are the effective mass and relaxation times along and across the optical axis, respectively.

The electromagnetic properties of the structure are described by Maxwell's equations

$$\text{rotH}=\frac{\partial}{\partial t}\text{D}, \quad \text{rotE}=-\frac{\partial}{\partial t}\text{B}, \tag{4}$$

and constitutive relations for bianisotropic medium

$$\text{D}=\varepsilon_0 \hat{\varepsilon}\,\text{E}-\frac{i}{c}\hat{\xi}\,\text{H}, \qquad \text{B}=\mu_0 \hat{\mu}\,\text{H}+\frac{i}{c}\hat{\xi}\,\text{E}, \tag{5}$$

where ε_0 and μ_0 are permittivity and permeability of free space, $c = (\varepsilon_0\mu_0)^{-1/2}$.

It is worth noting that in the research literature on chiral media there are several different constitutive relations in use [24], and the relations (4) are the most common and modern representation for bianisotropic media.

3. Partial Waves

3.1. Attenuation Constants

Theory of the surface electromagnetic waves reduces to finding solutions of Eqs. (4) in the form of inhomogeneous plane waves which decrease with increasing distance from the interface plane z=0: E,H ~exp[i(kr- ωt)], where the tangential components of the wave vector are real while the normal component is a complex quantity: $k_z = k_z^{'} + ik_z^{''}$. Therefore, for the region z>0 we have to choose solutions for which $k_z^{''}$>0 and for the region z<0 –the solutions with $k_z^{''} < 0$. Without losing generality, we can take the wave vector in the form k=(k $_x$,0, k $_z$). Then, using Eqs. (1-5) and excluding normal components of the fields

$$E_z = \frac{n}{\alpha_{\parallel}}(i\xi_{\parallel}E_y - \eta_0 H_y), \tag{6a}$$

$$H_z = \frac{n}{\alpha_{\parallel}}(i\xi_{\parallel}H_y + \eta_0^{-1}\varepsilon_{\parallel}E_y), \tag{6b}$$

for the L-material we obtain the following system of the wave equations:

$$i\xi_\perp E_x + \frac{k_z}{k_0} E_y + \eta_0 H_x = 0, \tag{7a}$$

$$\alpha_\| \frac{k_z}{k_0} E_x - i(\alpha_\| \xi_\perp + \xi_\| n^2) E_y - \eta_0 (\alpha_\| - n^2) H_y = 0, \tag{7b}$$

$$\varepsilon_\perp \eta_0^{-1} E_x - i\xi_\perp H_x - \frac{k_z}{k_0} H_y = 0, \tag{7c}$$

$$\varepsilon_\perp \eta_0^{-1} (\alpha_\| - \frac{\varepsilon_\|}{\varepsilon_\perp} n^2) E_y + \alpha_\| \frac{k_z}{k_0} H_x - i(\alpha_\| \xi_\perp + \xi_\| n^2) H_y = 0, \tag{7d}$$

where $n \equiv k_x / k_0$ is the index of the guiding surface modes,

$$\alpha_\| \equiv \varepsilon_\| - \xi_\|^2, \tag{8}$$

$k_0 = \omega / c$ is the wave number and $\eta_0 = \sqrt{\mu_0 / \varepsilon_0}$ is the intrinsic impedance of free space. The dispersion relation corresponding to Eqs.(7a-d) has a simple form

$$k_z^4 - 2Bk_0^2 k_z^2 + Ck_0^4 = 0, \tag{9a}$$

where

$$B = \varepsilon_\perp + \xi_\perp^2 + \alpha_\|^{-1} [\xi_\perp \xi_\| - (\varepsilon_\perp + \varepsilon_\|)/2] n^2, \tag{9b}$$

$$C = \alpha_\perp^2 - 2\frac{\alpha_\perp}{\alpha_\|} [\xi_\perp \xi_\| + (\varepsilon_\perp + \varepsilon_\|)/2] n^2 + \frac{\alpha_\perp}{\alpha_\|} n^4, \tag{9c}$$

$$\alpha_\perp \equiv \varepsilon_\perp - \xi_\perp^2. \tag{10}$$

Note that the dispersion relation for the R-material is the same as that given by Eq. (9a). In each medium, the solutions of Eq. (9a) describe two partial waves with different values for the normal component of the wave vector:

$$k_{z\pm}^2 = k_0^2 (B \pm \sqrt{B^2 - C}). \tag{11}$$

Therefore, the electric and magnetic field vectors in each medium are linear combination of the field vectors of both partial waves and can be written in the form (the common term $\exp[i(k_x x-\omega t)]$ is emitted):

$$E_L = E_+^L \exp(-\chi_+ k_0 z) + E_-^L \exp(-\chi_- k_0 z), \quad (12a)$$

$$H_L = H_+^L \exp(-\chi_+ k_0 z) + H_-^L \exp(-\chi_- k_0 z) \quad (12b)$$

in the region z>0, where $k_{z\pm} \equiv ik_0\chi_\mp$, and

$$E_R = E_+^R \exp(\chi_+ k_0 z) + E_-^R \exp(\chi_- k_0 z), \quad (13a)$$

$$H_R = H_+^R \exp(\chi_+ k_0 z) + H_-^R \exp(\chi_- k_0 z) \quad (13b)$$

for the R-material (z<0), where $k_{z\pm} \equiv -i\chi_\mp k_0$.

Using Eq. (11) we conclude immediately that in the absence of losses, $\mathrm{Im}\, k_z \neq 0$ and both partial waves in each medium are evanescent only if the following localization conditions are fulfilled simultaneously:

$$B<0, \quad 0<C<B^2. \quad (14)$$

Then both attenuation constants given by

$$\chi_\pm^2 = -B \pm \sqrt{B^2 - C} \quad (15)$$

are real and positive. In the case when $B>0,\ C>0$ and $C<B^2$, k_z has a real value and then both partial waves are propagating (bulk) waves. If $C<0$, the wave with k_{z+} is a bulk wave while that with k_{z-} is evanescent, but it loses energy via the propagating wave (so-called pseudosurface wave). And finally, for $C>B^2$ both k_{z+} and k_{z-} are complex quantities.

3.2. Field Structure in Partial Waves

The field structure in (+) and (-) partial waves can be described by the relations

$$E_\pm^L = \{\lambda_\pm, 1, i\vartheta_\pm\} E_{y\pm}^L,\ H_\pm^L = \{\rho_\pm, 1, i\psi_\pm\} H_{y\pm}^L, \quad (16a)$$

$$E_\pm^R = \{\lambda_\pm, 1, -i\vartheta_\pm\} E_{y\pm}^R,\ H_\pm^R = \{\rho_\pm, 1, -i\psi_\pm\} H_{y\pm}^R, \quad (16b)$$

where

$$\lambda_\pm = \frac{P_0\chi_\pm}{\alpha_\parallel\chi_\pm^2 + \alpha_\parallel\alpha_\perp - \alpha_\perp n^2}, \quad \vartheta_\pm = \frac{nP_\pm}{\alpha_\parallel\chi_\pm^2 + \alpha_\parallel\alpha_\perp - \alpha_\perp n^2}, \qquad (17)$$

$$P_\pm = \xi_\perp[(\xi-1)\chi_\pm^2 + \alpha_\perp(\xi+1)], \; Q_\pm = \xi_\perp[(\xi-\varepsilon)\chi_\pm^2 + \alpha_\perp(\xi+\varepsilon)], \quad (18)$$

$$\rho_\pm = \frac{Q_0\chi_\pm}{\alpha_\parallel\chi_\pm^2 + \alpha_\parallel\alpha_\perp - \alpha_\perp \varepsilon n^2}, \quad \psi_\pm = \frac{nQ_\pm}{\alpha_\parallel\chi_\pm^2 + \alpha_\parallel\alpha_\perp - \alpha_\perp \varepsilon n^2}, \quad (19)$$

$$P_0 = \xi_\perp[2\alpha_\parallel + (\xi-1)n^2], \; Q_0 = \xi_\perp[2\alpha_\parallel + (\xi-\varepsilon)n^2], \qquad (20)$$

$$\xi \equiv \xi_\parallel / \xi_\perp, \; \varepsilon \equiv \varepsilon_\parallel / \varepsilon_\perp. \qquad (21)$$

Note that in the absence of losses, the field components parallel to the interface plane are polarized linearly; moreover, they are parallel to each other:

$$E_{+x}^L / E_{+y}^L = E_{+x}^R / E_{+y}^R = \lambda_+, \; H_{+x}^L / H_{+y}^L = H_{+x}^R / H_{y+}^R = \rho_+ \quad (22)$$

and analogically for the (-) waves. Consequently, in this case the field vectors of each partial wave can be written as a sum of two vectors, one of which lies in the xy-plane, the other is parallel to the z-axis and shifted in phase by $\pi/2$ (or $-\pi/2$) with respect to the first vector. For example,

$$\mathrm{E}_\pm^L = \mathrm{E}_{\pm t}^L + \mathrm{E}_{\pm n}^L, \mathrm{E}_{\pm t}^L = \{\lambda_\pm, 1, 0\} \, \mathrm{E}_{\pm y}^L, \mathrm{E}_{\pm n}^L = i\{0, 0, \vartheta_\pm\} \, \mathrm{E}_{\pm y}^L. \qquad (23)$$

These two vectors represent the semiaxes of an ellipse described by the end point of vectors $\mathrm{E}_\pm^L$, respectively. The ellipsity (the ratio of the axes of the ellipse) is equal to $\sqrt{\lambda_+^2 + 1} / \vartheta_+$ for the (+) wave and $\sqrt{\lambda_-^2 + 1} / \vartheta_-$ for the (-) wave. It is evident that the field components parallel to the xz-plane (plane of propagation) are elliptically polarized too:

$$\frac{E_{\pm z}^L}{E_{\pm x}^L} = \left(\frac{E_{\pm z}^R}{E_{\pm x}^R}\right)^* = i\frac{nP_\pm}{P_0\chi_\pm}. \qquad (24)$$

4. DISPERSION RELATIONS OF THE SURFACE MODES

The dispersion relationship for surface waves, which establishes a connection between the frequency and tangential component of the wave vector, is obtained by using the standard boundary conditions for the tangential components of the fields at the interface plane z=0:

$$E^L_{+x} + E^L_{-x} = E^R_{+x} + E^R_{-x}, \quad H^L_{+x} + H^L_{-x} = H^R_{+x} + H^R_{-x}, \tag{25a}$$

$$E^L_{+y} + E^L_{-y} = E^R_{+y} + E^R_{-y}, \quad H^L_{+y} + H^L_{-y} = H^R_{+y} + H^R_{-y}. \tag{25b}$$

Using Eqs.(17-20) and eliminating x-components of the fields as well as the combinations ($E^L_{-y} - E^R_{-y}$) and ($H^L_{-y} - H^R_{-y}$), we obtain two distinct modes of surface plasmon polaritons which are described by the following dispersion relations:

$$n^2 = \alpha_{\parallel}[1 + \zeta(\alpha_{\parallel} - \alpha_{\perp})^{-1}] \equiv n_1^2, \tag{26}$$

$$n^2 = \varepsilon_{\perp}\alpha_{\parallel}(\varepsilon_{\perp}^2 + \varepsilon_{\parallel}\xi_{\perp}^2 + 2\varepsilon_{\perp}\xi_{\perp}\xi_{\parallel})(\alpha_{\parallel}\varepsilon_{\perp}^2 - \alpha_{\perp}\varepsilon_{\parallel}^2)^{-1} \equiv n_2^2, \tag{27}$$

where

$$\zeta \equiv (\xi_{\perp} + \xi_{\parallel})^2. \tag{28}$$

It is important to note that these modes exist only in the case of bianisotropic enantiomeric structure, for which $\varepsilon_{\perp} \neq \varepsilon_{\parallel}$ and $\xi_{\perp} \neq \xi_{\parallel}$. At a planar interface of two bi-isotropic enantiomeric media there are no surface waves [36].

In the surface mode described by Eq.(26), the tangential components of the electric fields in R- and L-materials are parallel, but $|E^L_{\pm t}| \neq |E^R_{\pm t}|$. Those for the magnetic field are parallel too, but $H^L_{\pm x} \neq -H^R_{\pm x}$ and $H^L_{\pm y} \neq -H^R_{\pm y}$. These inequalities are fulfilled, in particular, when $H^L_{\pm t} = H^R_{\pm t}$. In the second mode described by dispersion relation (27), we have $H^L_{\pm t} \parallel H^R_{\pm t}$ but $|H^L_{\pm t}| \neq |H^R_{\pm t}|$. As to the tangential components of the electric field, they may coincide:
$E^L_{\pm t} = E^R_{\pm t}$.

A comparison of Eqs. (26), (27) and dispersion relations for the surface electromagnetic waves localized at the interface between chiral and achiral media [9] as well as between two achiral media [5] shows that they are quite different.

5. SURFACE PLASMON POLARITONS IN NONDISSIPATIVE CASE

5.1. Resonance and Cutoff Frequencies

Before a rigorous examination of the waves' dispersion properties in general case of complex permittivity tensor components, it is useful to consider a simple situation, when there is no wave dissipation. In this case surface plasmon polaritons can be described like in the case of well known Fano modes [31] by real frequencies and real tangential wave vectors. Note also that in the lossless case ImB=ImC=0 [see Eqs.(9)] therefore the propagation of true surface waves with no space oscillations of the fields is possible.

Using the Drude expressions for the collisionless plasma

$$\varepsilon_{\perp,\parallel} = \varepsilon'_{\perp,\parallel} = 1 - \omega^2_{p\perp,\parallel} / \omega^2, \quad \varepsilon''_{\perp,\parallel} = 0, \tag{29}$$

the dispersion law (26) can be rewritten in the form

$$[n_1(\omega)]^2 = n^2_{1\infty} \frac{(\omega^2 - \omega_1^2)(\omega^2 - \omega_2^2)}{\omega^2(\omega^2 - \omega_r^2)} Sgn(\xi_\perp - \xi_\parallel), \tag{30}$$

where

$$n^2_{1\infty} = 2\xi_\perp(1 - \xi_\parallel^2) / |\xi_\perp - \xi_\parallel|, \tag{31}$$

$$\omega_1^2 = \omega^2_{p\parallel}(1 - \xi^2_\parallel)^{-1}, \quad \omega_2^2 = (\omega^2_{p\parallel} - \omega^2_{p\perp}) / \phi, \tag{32a}$$

$$\phi \equiv 2\xi_\perp(\xi_\perp + \xi_\parallel) \tag{32b}$$

and

$$\omega_r^2 = (\omega^2_{p\parallel} - \omega^2_{p\perp}) / (\xi^2_\perp - \xi^2_\parallel) \tag{33}$$

is the square of the resonance frequency. Analogically, the dispersion law (27) can be written in the form

$$n_2^2 = n^2_{2\infty} \frac{(\omega^2 - \omega_1^2)(\omega^2 - \omega^2_{p\perp})(\omega^2 - \varpi_+^2)(\omega^2 - \varpi_-^2)}{\omega^2(\omega^2 - \omega^2_{r1})(\omega^2 - \omega^2_{r2})(\omega^2 - \omega^2_{r3})} Sgn(\xi_\perp - \xi_\parallel), \tag{34}$$

where

$$n_{2\infty}^2 = (1+2\xi_\perp\xi_\parallel + \xi_\perp^2)(1-\xi_\parallel^2)/\,|\,\xi_\perp^2 - \xi_\parallel^2\,|, \tag{35}$$

$$\varpi_\pm^2 = \omega_{p\perp}^2\{1+\xi_\perp\xi_\parallel + \gamma_p\xi_\perp^2/2 \pm \xi_\perp\left[(1+\xi_\perp\xi_\parallel)\gamma_p + \gamma_p^2\xi_\perp^2/4 - 1 + \xi_\parallel^2\right]^{1/2}\}(1+\zeta-\xi_\parallel^2)^{-1}, \tag{36}$$

$$\gamma_p \equiv \left(\omega_{p\parallel}/\omega_{p\perp}\right)^2 = m_\perp / m_\parallel \tag{37}$$

is the anisotropy parameter and the resonance frequencies $\omega_{r1,2,3}$ are given by solutions of the equation bicubic in ω:

$$\omega^6 + \omega_r^2(\omega^2-\omega_{p\parallel}^2)(\omega^2-\omega_{p\perp}^2) + 2\omega^2\omega_{p\perp}^2(\xi_\perp^2-\xi_\parallel^2)^{-1}[(\xi_\parallel^2-\gamma_p\xi_\perp^2)\omega^2 - \omega_{p\perp}^2(\xi_\parallel^2-\gamma_p^2\xi_\perp^2)/2] = 0. \tag{38}$$

Note that the waves governed by dispersion relations (30) and (34) can be considered as a novel class of Fano modes caused by bianisotropy on a planar interface between enantiomeric plasmonic media with identical permittivities.

Furthermore, for the brevity we will restrict us to the consideration of the mode described by relation (30). Properties of the second mode governed by Eqs.(34) and can be examined analogically.

It is clear from Eq.(30) that the wave is cut off (k_x =0) at $\omega = \omega_1$ and $\omega = \omega_2$, if $\gamma_p > 1$, and only at $\omega = \omega_1$, if $\gamma_p < 1$. Moreover, the wave has a resonance ($k_x \to \infty$) only if the expression on the right hand side of Eq.(33) is positive. It means that the surface plasmon modes with frequency $\omega = \omega_r$ in the absence of retardation occur only when $\gamma_p > 1$ if $\xi_\parallel < \xi_\perp$, and vice versa: only at $\gamma_p < 1$, if $\xi_\parallel > \xi_\perp$.

Note that at the resonance frequency $\varepsilon_\parallel(\omega_r) - \varepsilon_\perp(\omega_r) = \xi_\parallel^2 - \xi_\perp^2$ while at cutoff frequencies $\varepsilon_\parallel(\omega_1) = \xi_\parallel^2$ and $\varepsilon_\parallel(\omega_2) - \varepsilon_\perp(\omega_2) = -\phi$. Localization conditions (14) give

$$\alpha_\parallel > \alpha_\perp > -(\varepsilon_\parallel - \varepsilon_\perp)^2/4\zeta, \tag{39}$$

therefore the condition $\chi_+\chi_- = \alpha_\perp\zeta(\alpha_\perp - \alpha_\parallel)^{-1} > 0$ can only be fulfilled if $\alpha_\perp$ is negative.

5.2. Frequency Regions of the Wave Existence

Using left condition (39), one can conclude that surface plasmon polaritons exist in the frequency range

$$\omega_r < \omega < \omega_3 \equiv \omega_{p\perp}(1-\xi_\perp^2)^{-1/2} \quad (40)$$

if $\xi_\parallel < \xi_\perp$, and in the region

$$\omega < \min\{\omega_3, \omega_r\} \quad (41)$$

if $\xi_\parallel > \xi_\perp$. However, to find the precise regions of existence, we have to take into account the unequality on the right hand side of Eq.(39) too. Before to do that, it is convenient to introduce two additional definitions:

$$\omega_4 = \omega_{p\parallel}(1-\xi_\perp^2-\zeta)^{-1/2}, \quad \omega_5 = \omega_{p\parallel}[1-\xi_\parallel^2+(\xi_\parallel-\xi_\perp)^2/4]^{-1/2}. \quad (42)$$

Note that

$$\varepsilon_\parallel(\omega_4) = \varepsilon_\perp(\omega_3)+\zeta = \zeta+\xi_\perp^2, \quad \varepsilon_\parallel(\omega_5) = \xi_\parallel^2-(\xi_\parallel-\xi_\perp)^2/4. \quad (43)$$

In Figure 1 the region in the plane $(\alpha_\parallel, \alpha_\perp)$ is shown where conditions (39) are satisfied and both attenuation constants χ_+ and χ_- are positive (shaded). It is easily to see that for a given value of $\alpha_\parallel$ in the interval $-(\xi_\parallel-\xi_\perp)^2/4 < \alpha_\parallel < \phi$, that is, inside the frequency range

$$\omega_5 < \omega < \omega_4, \quad (44)$$

there is a forbidden gap for $\alpha_\perp$ between α_+ and α_-:

$$\alpha_- < \alpha_\perp < \alpha_+, \quad (45)$$

where

$$\alpha_\pm(\alpha_\parallel) = -[\xi_\perp+\xi_\parallel \mp \sqrt{\phi-\alpha_\parallel}]^2. \quad (46)$$

The corresponding frequency forbidden gap is given by

$$\omega_- < \omega < \omega_+, \quad (47a)$$

where

$$\omega_{\pm}^2 = \omega_3^2\{1 \pm [1-(1-\gamma)^2\zeta^{-1}(1-\xi_{\perp}^2)]^{1/2}\}/2. \tag{47b}$$

With increasing $\alpha_{\|}$, the forbidden gap decreases and vanishes at $\alpha_{\|} = \phi$. It means that for $\alpha_{\|} \geq \phi$, the waves exist at any negative value of $\alpha_{\perp}$. The corresponding frequency range is given by

$$\omega_4 \leq \omega < \omega_3 \tag{48}$$

and occurs only if the anisotropy parameter

$$\gamma_p < \gamma_2 \equiv 1-\zeta(1-\xi_{\perp}^2)^{-1}. \tag{49}$$

with decreasing $\alpha_{\|}$ in the region $\alpha_{\|} \leq -(\xi_{\|}-\xi_{\perp})^2/4$, the forbidden gap is expanded and can be represented by unequality $\alpha_{-}(\alpha_{\|}) < \alpha_{\perp} < \alpha_{\|}$, where $\alpha_{-}(\alpha_{\|})$ is given by Eq. (46). Using now Eqs.(30) and (39), it is easily to see that the wave can only exist ($n^2>0$) if $\alpha_{\|} > 0$, that is, in the frequency range

$$\omega_1 < \omega < \omega_3, \tag{50}$$

and this is only possible if

$$\gamma_p < \gamma_0 \equiv (1-\xi_{\|}^2)(1-\xi_{\perp}^2)^{-1}. \tag{51}$$

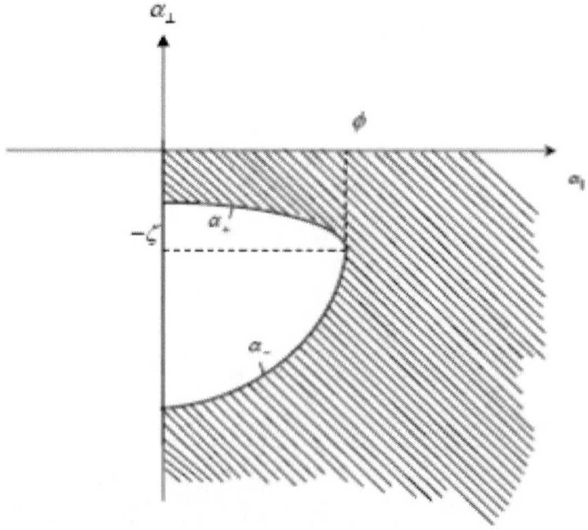

Figure 1. The region in the plane $(\alpha_{\|}, \alpha_{\perp})$ where the conditions (18) are satisfied and both χ_{+} and χ_{-} are real positive quantities (shaded).

6. Dispersion Curves

The dispersion curves of lossless surface plasmon polaritons are shown schematically in Figures 2 and 3 for different values of the ratio $\xi = \xi_{\|} / \xi_{\perp}$ and different intervals of the anisotropy parameter γ_p. In the case when $\xi < 1$ and $1 < \gamma_p < \gamma_0$, there are two branches for which $n^2 > 0$: low-frequency branch corresponding to the propagating (bulk) waves (dashed lines in Figure 2a) and higher-frequency branch, only two sections of which correspond to the surface waves of interest (solid lines). There is a forbidden bandgap given by Eq.(47a), so as the wave exists only in the following two frequency ranges:

$$\omega_1 < \omega < \omega_-, \qquad \omega_+ < \omega < \omega_3. \tag{52}$$

In the case of $\gamma_1 < \gamma_p < 1$, where

$$\gamma_1 = 1 - \zeta^{1/2}(1 - \xi_\perp^2)^{-1/2}, \tag{53}$$

both ω_r^2 and ω_2^2 are negative. It means that low-frequency branch on Figure 2a is absent while the sections of the high-frequency dispersion curve corresponding to the surface wave are the same. Note that in the case $\gamma_p \le \gamma_1$ (Figure 2b), the forbidden gap disappears and the wave exists in all frequency range given by Eq. (50). If $\gamma_p = \xi = 1$ (the case of bi-isotropic media), as it has been mentioned yet, there are no surface waves. Dispersion curves in the case $\xi > 1$ are shown in Figs. 3a,b, for two different intervals of the parameter γ_p.

In the case of Figure 3a, when

$$\gamma_1 < \gamma_p \le \gamma_2, \tag{54}$$

there is again a forbidden gap and the wave exists in the same two frequency ranges which are given in Eq.(52). In the case $\gamma_p \le \gamma_1$ (Figure 3b) the forbidden gap disappears, but unlike to the case of Figure 2b, when the curve approaches asymptotically to the straight line $\omega = ck_x / n_\infty$ with increasing k_x, now it approaches to the resonance line $\omega = \omega_r$.

a. $\gamma_1 < \gamma_p \le \gamma_2$. b. $\gamma_p \le \gamma_1$.

Thus, we conclude that the dispersion properties of the surface waves are very sensitive to the value of the ratio $\xi_{\|} / \xi_{\perp}$ as well as to the value of the anisotropy parameter γ_p given by Eq.(37).

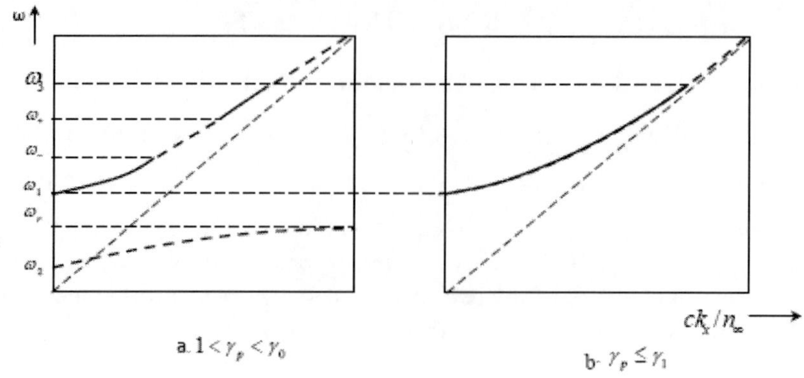

Figure 2. Dispersion curves of SPP (solid lines) in the case $\xi_\parallel < \xi_\perp$. (Dashed lines correspond to the bulk waves.)

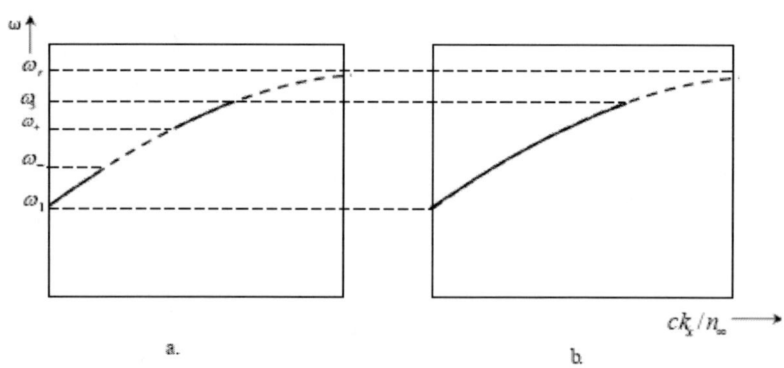

Figure 3. Dispersion curves of the surface modes (solid lines) in the case $\xi_\parallel > \xi_\perp$.

7. Refraction Index in the Presence of the Wave Dissipation

Consider now briefly the main properties of surface plasmon polaritons governed by Eq. (26) taking damping into account. Substituting Eqs.(3a,b) into (26) and separating real and imaginary parts $(n = n' + in'')$ we obtain

$$n'^2 = \left(F + \sqrt{F^2 + G^2}\right)/2, \quad n'' = G/2n', \tag{55}$$

where

$$F = \alpha'_\parallel + \zeta F_1/F_0, \quad G = \varepsilon''_\parallel + \zeta G_0/F_0, \quad F_0 = (\alpha'_\parallel - \alpha'_\perp)^2 + (\varepsilon''_\parallel - \varepsilon''_\perp)^2, \tag{56}$$

$$F_1 = \alpha_{\|}^{'}(\alpha_{\|}^{'} - \alpha_{\perp}^{'}) + \varepsilon_{\|}^{''}(\varepsilon_{\|}^{''} - \varepsilon_{\perp}^{''}), \quad G_0 = \alpha_{\|}^{'}\varepsilon_{\perp}^{''} - \alpha_{\perp}^{'}\varepsilon_{\|}^{''},$$

$$\alpha_{\|,\perp}^{'} = \varepsilon_{\|,\perp}^{'} - \xi_{\|,\perp}^{2}. \qquad (57)$$

The relative attenuation of the wave on the wavelength distance is given by

$$n^{''} / n^{'} = \left(\sqrt{F^2 + G^2} - F\right) / G. \qquad (58)$$

Since the square of the refraction index defined by Eq.(55) is always positive independly on the sign of F, it is clear that the wave can propagate not only in the frequency regions defined in the previous section, but also at any other frequencies at which the dissipation is not very strong. Besides, the presence of losses leads to a finite value of the refraction index at resonance frequency ω_r, as well as to nonzero values at cutoff frequencies $\omega_{1,2}$ given by Eqs. (32a,b). Indeed,

$$n'^2(\omega_r) = \left(\alpha_{\|}^{'} + \zeta\varepsilon_{\|}^{''} / \delta + \sqrt{(\alpha_{\|}^{'2} + \varepsilon_{\|}^{''2})(1 + \zeta^2 / \delta^2)}\right) / 2, \qquad (59a)$$

$$n'^2(\omega_1) = \varepsilon_{\|}^{''}\left(\zeta\delta + \sqrt{(\alpha_{\perp}^{'2} + \delta^2)[\delta^2 + (\alpha_{\perp}^{'} - \zeta)^2]}\right) / 2(\alpha_{\perp}^{'2} + \delta^2), \qquad (59b)$$

$$n'^2(\omega_2) = \delta\left(\alpha_{\|}^{'}\delta + \zeta\varepsilon_{\|}^{''} + \sqrt{(\alpha_{\|}^{'2} + \varepsilon_{\|}^{''2})(\delta^2 + \zeta^2)}\right) / 2(\zeta^2 + \delta^2), \qquad (59c)$$

where $\delta = \varepsilon_{\|}^{''} - \varepsilon_{\perp}^{''}$ and the values of $\alpha_{\|,\perp}^{'}, \varepsilon_{\|,\perp}^{''}$, are assumed to be taken at corresponding frequencies in each of Eqs.(59a-c).

8. PROPAGATION LENGTH

One of the most important characteristics of surface waves in lossy structures is the propagation length defined as a distance along the propagation direction where the mode power decays by a factor of 1/e. This distance can be calculated using the expression

$$L \equiv (2\,\mathrm{Im}\,k_x)^{-1} = n^{'} / k_0 G. \qquad (60)$$

It is evident that in the absence of damping $L \to \infty$ while the absorption leads to its finite value depending on ω. Note that all the expressions given by Eqs.(55-58) and (60) are highly general and applicable for all frequencies at any level of dissipation. Assuming that the dissipation is weak $(n^{''} / n^{'} << 1)$, in Eq. (55) one can put $n^{'} \approx \sqrt{|F|}$, at this the extinction coefficient $n'' \approx G/2\sqrt{|F|}$. Note also that in the limiting case $\omega >> \tau_{\|,\perp}^{-1}$ (for example, at visible and shorter wavelengths), the imaginary parts of the permittivity tensor components can be approximated as

$$\varepsilon''_{\|,\perp} \approx \frac{\omega^2_{p\|,\perp}}{\omega^2(\omega\tau_{\|,\perp})} \tag{61}$$

while the real parts are given by Eq.(29). At this end, for frequencies far enough from the plasma frequencies we have

$$|\varepsilon'_{\|,\perp}| >> \varepsilon''_{\|,\perp}, \ |\omega^2 - \omega^2_{p\|,\perp}|/\omega^2_{p\|,\perp} >> (\omega\tau_{\|,\perp})^{-1}. \tag{62}$$

Then in the first nonvanishing approximation over $1/\varepsilon''_{\|,\perp}$ one can obtain from Eq.(60)

$$L(\omega) = n_1(\omega)/k_0[\varepsilon''_{\|} + \zeta G_0(\alpha'_{\|} - \alpha'_{\perp})^{-2}], \tag{63}$$

where $n_1(\omega)$ is given by Eq.(30). However, Eq. (63) is not applicable for frequencies close to the resonance and cutoff frequencies. Assuming that $|\delta|/\zeta << 1$ and using Eqs.(59a-c) and (60), it is easily to see that in the vicinity of these characteristic frequencies Eq.(63) should be replaced by

$$L(\omega_r) \approx k_0^{-1}\sqrt{|\delta|/2\zeta|\alpha'_{\perp}|}, \tag{64a}$$

$$L(\omega_1) \approx k_0^{-1}\sqrt{|\alpha'_{\perp}|/2\varepsilon''_{\|}|\alpha'_{\perp} - \zeta|}, \tag{64b}$$

$$L(\omega_2) \approx k_0^{-1}\sqrt{\zeta|/2|\alpha'_{\|}\delta|}, \tag{64c}$$

respectively. According to Eqs.(64a-c), the propagation length in the vicinity of cutoff frequencies, unlike that for the resonance frequency, becomes large enough for a possibility of an experimenrtal observation of the surface mode. To excite the surface wave at an interface between enantiomeric bianisotropic plasmonic media, the well known attenuated total reflection method and Otto-Kretchmann configuration[3] can be used.

CONCLUSION

The possibility of existence of surface plasmon polaritons at the interface between two enantiomorph uniaxial bianisotropic materials has been shown. The waves are caused by bianisotropy and are absent in the case of enantiomorph isotropic chiral plasmonic materials. Such waves exist under special conditions in various frequency regions. Field structure and polarization properties of the partial waves are studied. The existence of two distinct branches described by various dispersion relations and localization conditions is shown. Frequency regions of existence, as well as cutoff and resonance frequencies of the waves are found, localization conditions and dispersion curves in the absence of dissipation are examined. The effect of losses on the wave's refractive index is investigated and the propagation length is determined for frequencies far enough from the plasma frequencies as well as in the vicinity of the characteristic frequencies. It is shown that for visible and shorter wavelengths, the propagation length in the vicinity of cutoff frequencies becomes large enough for the observation of the surface plasmon polaritons.

The dispersion relations and localization conditions differ dramatically from those for the conventional surface electromagnetic waves propagating along the interface between chiral and achiral media, as well as between two achiral media with various dielectric permittivities. This difference leads to the unique properties of the considered surface waves which can be used for wide range of potential applications including biosensing.

References

[1] A. Sommerfeld. *Annalen der Physic und Chemie*. 67 (1899) 233-290.
[2] H. Barlow and J. Brawn. *Radio Surface waves*. N.Y., Oxford, 1962.
[3] A. Otto. *Z.Phys*. 216 (1968) 398.
[4] D.L. Mills, E.Burstein. *Rep. Prog. Phys*. 37 (1974) 817.
[5] G. Borstel, H.Falge and A. Otto. *Springer Tracts Modern Physics* 74 (1974) 107.
[6] R. H. Tarkhanyan. *Phys. Stat. Solidi* (b) 72 (1975) 111-122.
[7] V.M. Agranovich and D.L. Mills (eds.). Surface polaritons. North-Holland, Amsterdam, 1982.
[8] A.D. Boardman (ed.) Electromagnetic Surface Modes. Wiley and Sons, 1982.
[9] N. Engheta and P. *Pelet. Opt. Lett*. 16 (1991) 723-725.
[10] S. Kawata (ed.) Near-field optics and surface plasmon polaritons. Springer, 2001.
[11] V.I. Alshits and V.N. Lyubimov. *Phys. Solid State* 44 (2002) 1988-1992.
[12] V.L. Barnes, A. Dereux. T.W. Ebbesen. *Nature* (London) 424 (2003) 824.
[13] N. Kedei et al. *Canser Des*. 64 (2004) 3243.
[14] I. Abdulhalim, M. Zourob, A. Lakhtakia. *Electromagnetics* 28 (2008) 214-242.
[15] J.B. Pendry. *Phys. Rev. Lett*. 85 (2000) 3966-3969.
[16] R. A. Shelbi, D.R. Smith, S. Shultz. *Science* 292 (2001) 77.
[17] S. Tretyakov, A. Sihvola, L. Jylha. *Photonics Nanostr. Fundam. Appl*. 3 (2005) 107.
[18] R. Moussa, Th. Koschny and C.M. Soukoulis. *Phys. Rev. B* 74 (2006) 115111.
[19] C.M. Soukoulis, S. Linden, M. Vegener. *Science* 315 (2007) 47.
[20] E. Plum, J. Zhou, J. Dong, V. Fedotov, T. Koschny, C. Soukoulis, N. Zheludev. *Phys. Rev. B* 79 (2009) 035407.
[21] R.H. Tarkhanyan, D.G. Niarchos, M. Kafesaki. *J. Magn. Magn. Mater*. 322 (2010) 603-608.
[22] R.Ruppin. *J. Phys.: Condens. Matter* 13(2001) 1811.

[23] J. Bravo-Abad, F. Garsca-Vidal and L. Martin-Moreno. *Phys. Rev. Lett.* 93 (2004) 227401.
[24] A. Lakhtakia, V.V, Varadan, V.K. Varadan. *J. Opt. Soc. Amer. A* 5 (1988) 175-184.
[25] S. Tretyakov, I. Nefedov, A. Sihvola, S. Maslovski, C. Simovski. J. Electromagn. *Waves Appl.* 17 (2003) 695-706.
[26] J. B. Pendry. *Science* 306 (2004) 1353-1355.
[27] A. Boardman, N. King, Yu. Rapoport, L. Velasco. *New J. Phys.* 7 (2005) 191.
[28] C. Qiu, H. Yao, L. Li, S. Zouhdi, T. Yeo. *Phys. Rev. B* 75 (2007) 155120.
[29] B. Vang, J. Zhou, T. Koschny, M. Kafesaki, C. Soukoulis. *J. Opt. A: Pure Appl. Opt.* 11 (2009) 114003.
[30] J. Zenneck. *Ann. Phys. Lpz*. 23 (1907) 846-866.
[31] U. Fano. *J. Opt. Soc. Amer*. 31 (1941) 213-222.
[32] H. Raether. Surface Plasmons. Springer-Verlag, Berlin,1988
[33] M.I. Dyakonov. *Sov. Phys. JETP* 67 (1988) 714-716.
[34] O. Takayama, L. Crasovan, et al. *Phys. Rev. Lett*. 102(2009) 043903 .
[35] J. Gao, A. Lakhtakia, M. Lei, *Phys. Rev.A* 81 (2010) 013801.
[36] R. H. Tarkhanyan, N.G. Niarchos, *Phys. Status Solidi B*, 248 (2011) 1499-1504.
[37] N.V. Ashcroft, N.D. Mermin, *Solid State Physics* (Brooks Cole, 1976).

In: Plasmons: Structure, Properties and Applications
Editors: Anton E. Turunen and Johannes O. Niemi
ISBN: 978-1-62100-060-0

Chapter 8

Raman Studies of Anisotropy of Phonon-Plasmon Modes in GaAs/AlAs Doped Superlattices Grown on Flat and Nanofaceted Surfaces

V. A. Volodin*

A.V. Rzhanov Institute of Semiconductor Physics,
Siberian Branch of Russian Academy of Sciences, Novosibirsk, Russia
Novosibirsk State University, Novosibirsk, Russia

Abstract

Raman spectroscopy technique was applied to study phonon-plasmon coupled modes in Si doped set of GaAs/AlAs superlattices (SLs), grown on (100), (311)A and (311)B substrates. Anisotropy in dispersion of phonon-plasmon modes with wave-vector normal to SL layers and in SL layers was observed for (100) SLs. Additional lateral anisotropy of the phonon-plasmon modes was observed for SLs grown on nano-corrugated (311)A surface. The observed anisotropy is supposed to be result of charge carrier mass anisotropy appearing due to structural anisotropy.

Keywords: GaAs/AlAs superlattices, plasmons, Raman scattering, phonon-plasmon coupling, corrugation, anisotropy

Introduction

Effects of interaction of phonons with electrons in bulk semiconductors and semiconductor superlattices (SLs) have long been under study of wide community of researchers [1] but remain actual up today. The analysis of dispersion of phonon-plasmon

* E-mail address volodin@isp.nsc.ru

modes in SLs can be express and non-destructive tools for measurements of charge carrier concentration and effective mass anisotropy [2,3].

According to calculations [2,3], in the case of the effective mass anisotropy of charge carriers, the phonon–plasmon modes propagating in the directions of “heavy” and “light” masses should have different frequencies. In this case, the mass difference is not as important as the wave-number of the phonon–plasmon modes. For the frequency difference to be experimentally observable (e.g., at least a few inverse centimeters), the wave-number of the modes must be on the order of one million of reversed centimeters. Since the semiconductor SLs are optically opaque, they may be Raman studied only using backscattering geometry. To reach a necessary wave-number difference, it is insufficient to only vary the angle of incidence of light relative to normal of a SL. It is necessary to implement the normal incidence geometry and the geometry in which the wave vectors of incident and scattered light lie in a SL plane. This is possible if the backscattering occurs at the side face of the SL as shown in Figure 1. Since usually the thickness of the SLs is less than a micron, one has to use a Raman scattering setup with micro-Raman setup.

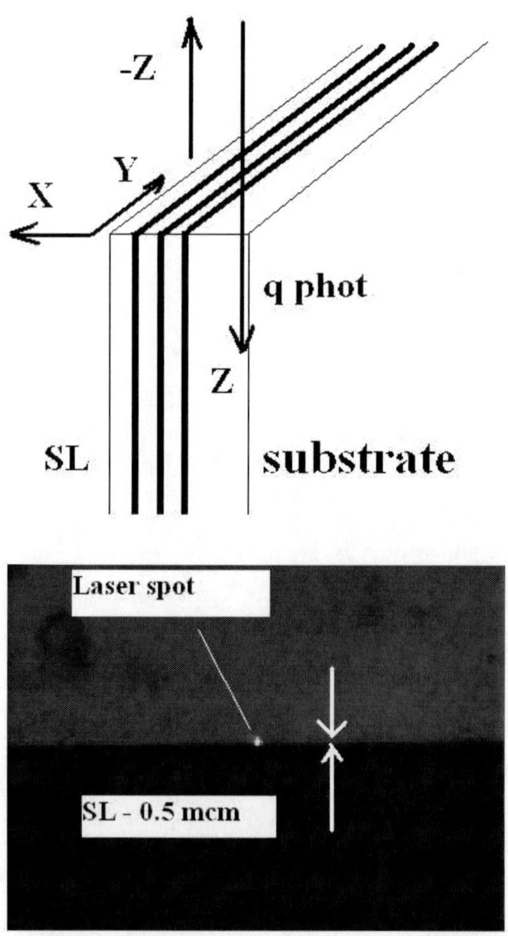

Figure 1. Scheme of “in-plane” Raman geometry (up) and micro-photography of edge of a sample with a laser spot (bottom).

EXPERIMENTAL

The studied SLs were grown using molecular beam epitaxy technique on (311)A-(1x8) reconstructed GaAs substrate and (001)-(2x4) reconstructed GaAs substrate at temperature 550° C. The control of reconstruction was carried out *in situ* with the use of reflection of high energy electron diffraction. In the case of (311) substrates the angle disorientation of the substrates from (311) direction was less than 15'.The flux of As_4 molecules was $5*10^{14}$ molecules/cm^2 per second, the growth rate was about 0.3 μm/h for GaAs and 0.32 μm/h for AlAs. All SL-satellites on (001) and (311)B GaAs substrates were grown in the same growth procedures – "side by side".

The (311)A and B surfaces were detected using anisotropy of chemically etching. The condition of growth are more detailed subscribed elsewhere [4]. The set of $GaAs_n/AlAs_m$ SLs were grown, data are shown in Table 1. To achieve electron concentration at room temperature $2*10^{18}$ cm^{-3}, the impurity (Si) concentration in doped SLs was $2.5*10^{19}$ cm^{-3}. Impurity of Si can be both acceptor and donor, but in chosen growth condition it give the n-type material.

Table 1. Specifications of the studied SLs $GaAs_nAlAs_8$. 1 ML in 311 direction = 1.7 angstrom

Substrate			GaAs thickness, Å.	AlAs thickness, Å.	Periods
311A	311B	100			
1A	1B	1	1.7	13.6	400
2A	2B	2	3.4	13.6	300
3A	3B	3	6.8	13.6	200
4A	4B	4	10.2	13.6	100

The Raman spectra were registered at room temperature in backscattering geometry in various polarization geometries. The line 514.5 nm of Ar^+ laser was used as a light source. The triple spectrometer T64000 (Horiba Jobin Yvon) equipped with cooled CCD camera and micro-Raman setup was used. Polarization resolved Raman spectra were registered using polarization optic elements.

RESULTS AND DISCUSSION

It is known, that in polar semiconductors the phonon-plasmon interaction takes place. It leads to coupling of phonon-plasmon modes and to shift and broadening of Raman peaks [1]. The using of micro-Raman system allows to focus laser beam into SL edge in sport about 1 micrometer (Figure 1). The Raman spectra of (100) GaAs/AlAs SLs for "normal" geometry are shown in figure 2.

One can see shift of LO^+ phonon-plasmon modes versus variation of GaAs layer thickness. The peaks at 240-280 cm^{-1} are due to scattering on localized optic phonons in GaAs layers. One can see also longitudinal optic (LO) bulk modes from GaAs substrate. It should be noted, that scattering on transverse optical (TO) phonons is forbidden in this geometry, but its appear supposal due to resonance phenomena.

In figure 3 one can see Raman spectra of (100) SLs registered in Z(YY)-Z geometry, Wave-vector of initial light was parallel to Z axe, wave-vector of scattered light was anti-parallel to Z axe, vector of polarization of light was parallel to SL layers (see figure 1). In the case of (100) SLs axe X is along (100), axe Y is along (01-1), and axe Z is along (011) directions.

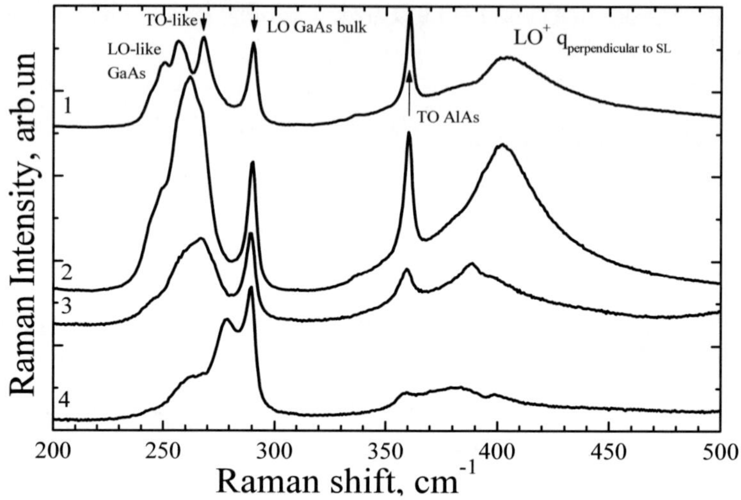

Figure 2. Raman spectra in “normal” geometry for (100) SLs, curves 1-4 correspond to samples 1-4 in Table 1.

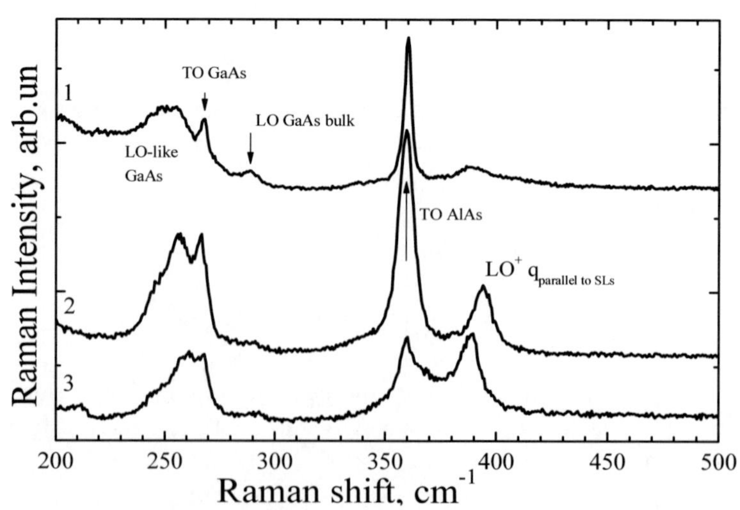

Figure 3. Raman spectra in “in-plane” geometry for (100) SLs 1-3.

It should be noted, that the spectra in this “in-plane” geometry practically do not depend on polarization of light (the configurations Z(XX)-Z and Z(XY)-Z were also used). If one compare the spectra of the same (100) SLs but in various geometry (figures 2 and 3), the anisotropy of phonon-plasmon modes is clearly seen. The wave-vector is perpendicular to SL layers in figure 2 and wave-vector is parallel to SL layers in figure 2.

It should be noted, that all SLs with the used thickness ratio of GaAs and AlAs must be of type II. The mixing of electron states of the Γ and *X* valleys occurs upon the transition of the GaAs/AlAs superlattice from type I to type II. The electron effective mass in the *X* valleys of both GaAs and AlAs is anisotropic, and the artificial symmetry of the superlattice period contributes to the mass anisotropy. Additional lateral anisotropy can be due to lateral nanofaceting of surface during SL growth.

In figure 4 one can see an example of lateral anisotropy of (311)A SLs. It is known, that in some growth condition in (311)A surface GaAs form quantum well wire (QWW) like structure [5]. A model of growth of QWW structure is shown in figure 4.

The (311)A surface is nano-corrugated with lateral period 3.2 nm along (01-1) direction. So, it is possible to form structure with QWWs elongated along (-233) direction. In the case of (311) SLs axe X' is along (01-1), axe Y' is along (-233), and axe Z' is along (311) directions. Solid line in figure 5 corresponds to Z'(X'X')-Z' geometry, dashed line corresponds to Y'(X'X')-Y' geometry, dot line corresponds to X'(Z'Z')-X' geometry. It should be noted, that axes X', Y', and Z' do not correspond to axes X, Y, and Z shown in figure 1.

One can see in figure 5 anisotropy for "in-plane" geometry. Wavevector can be parallel to QWWs (dashed line) or wavevector can be perpendicular to QWWs (dot line) in figure 5. In the last case the spectrum more like on spectrum in "normal" geometry (wavevector is also perpendicular to QWWs - solid line). It should be noted, that in similar (311)A SL optical anisotropy of light absorbance (dichroizm) was observed [6].

It is known, that structural lateral anisotropy is not so pronounced in the case of (311)B SLs [7]. The Raman experiment was also carried out for 2B SLs (see table 1). Solid line in figure 6 corresponds to Z'(X'X')-Z' geometry, dashed line corresponds to Y'(X'X')-Y' geometry. One can see anisotropy for "normal" and "in-plane" geometries. But for lateral effects, SLs (311)B were more like on (100) SLs without pronounced lateral anisotropy.

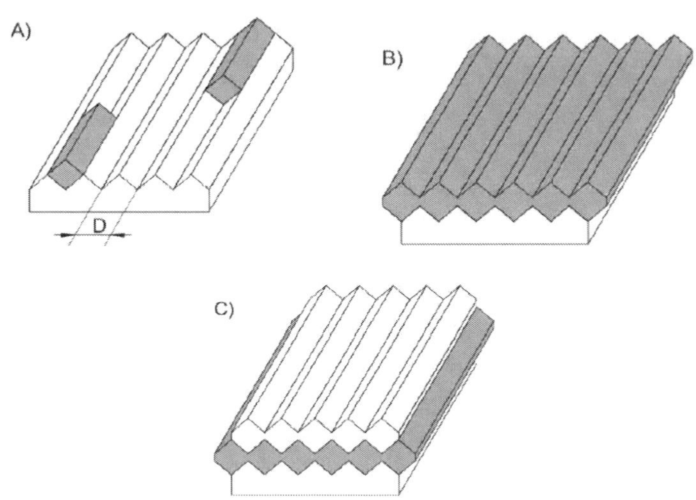

Figure 4. Model of growth of anisotropic quantum-well-wire like SLs on a corrugated surface. A – partial covering; B – full covering; C – covering by other material.

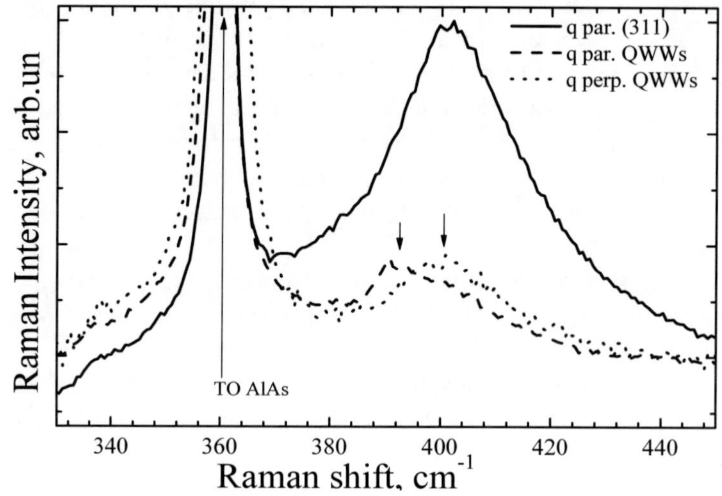

Figure 5. Raman spectra in "normal" and"in-plane" geometries for 311A SL (sample 2A – see Table 1). Arrows show the lateral anisotropy effect.

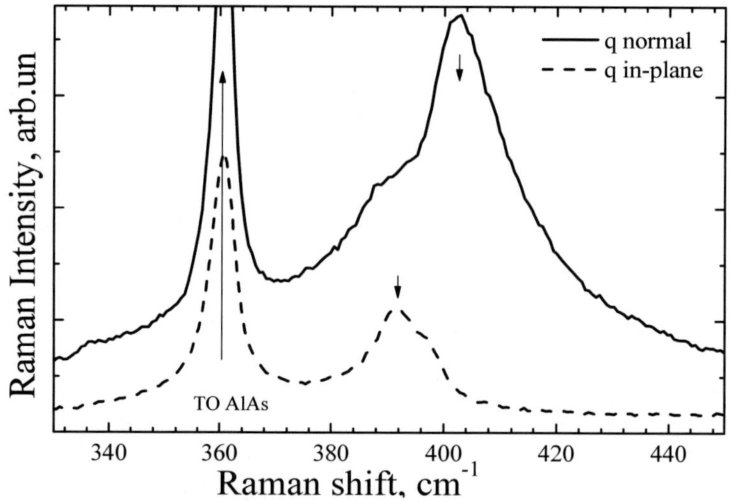

Figure 6. Raman spectra in "normal" and "in-plane" geometries for 311B SL (sample 2B – see Table 1). Arrows show the anisotropy effect.

SUMMARY

So, as it was predicted in calculations [2,3], the anisotropy in dispersion of phonon-plasmon modes, supposed due to effective mass anisotropy, was observed. The lateral anisotropy for (311)A SLs grown on nano-corrugated surfaces was observed.

ACKNOWLEDGMENTS

Author is grateful to Drs. Preobrazhenskiy and Semyagin for growth of the SLs.

REFERENCES

[1] Light Scattering in Solids IV.Electronic Scattering, Spin Effects, SERC, and Morphic Effects, edited by Cardona M and Günterodt G (Springer-Verlag, Berlin, 1982).

[2] V.A.Volodin, M.D.Efremov, V.A.Sachkov, N.N.Ledentsov. *Phys. Low-Dim. Strut.* 5/6 (2003), p. 109.

[3] Nanophysics, Nanoclusters and Nanodevices: Phonon-plasmon coupling modes in tunneling thin GaAs/AlAs (311) and (001) SLs: Raman studies and modeling, edited by Kimberly S. Gehar (Nova Science Publishers Inc., New York, 2006).

[4] Trends in Nanotechnology Research: Interface reconstruction in GaAs/AlAs ultrathin superlattices grown on (311) and (001) surfaces, edited by Eugene V. Dirote (Nova Science Publishers Inc., New York, 2004).

[5] R. Nötzel, N.N. Ledentsov, L.A. Däweritz, M. Hohenstein and K. Ploog, *Phys. Rev. Lett.* 67 (1991), p. 3812.

[6] V.A. Volodin, M.D. Efremov, R.S. Matvienko, V.V. Preobrazhenskii, B.R. Semyagin, N.N. Ledentsov, I.R. Soshnikov, D. Litvinov, A. Rosenauer, and D. Gerthsen. *Physics of the Solid State*, 47, (2005), p. 366.

[7] N.N.Ledentsov, D.Litvinov, A.Rosenauer, D.Gerthsen, I.P.Soshnikov, V.A.Shchukin, V.M.Ustinov, A.Yu.Egorov, A.E.Zukov, V.A.Volodin, M.D.Efremov, V.V.Preobrazhenskii, B.P.Semyagin, D.Bimberg and Zh.I.Alferov. *Journal of Electronic Materials*, v. 30(5) (2001), p. 463.

INDEX

C

D

E

F

G

H

I

K

L

M

N

O

P

Q

R

S

T

U

V

W

X

Y